NONLINEAR DYNAMICS IN OPTICAL COMPLEX SYSTEMS

Advances in Optoelectronics (ADOP)

Editor-in-chief: Takeshi Kamiya (University of Tokyo)

Associate Editor: Motoichi Ohtsu (Tokyo Institute of Technology)

Editorial Board Members:

    John E. Bowers (University of California at Santa Barbara)
    Daniel Courjon (Institut des Microtechniques de Franche-Comte)
    Ernst Goebel (Physikalisch Technische Bundesanstalt)
    Tatsuo Izawa (NTT Electronics)
    Satoshi Kawata (Osaka University)
    Kazuroh Kikuchi (Univerisity of Tokyo)
    David A. M. Miller (Stanford University)
    David N. Payne (University of Southampton)
    Klaus Petermann (Technical University of Berlin)
    Yasuharu Suematsu (Kochi University of Technology)
    Yoshihisa Yamamoto (Stanford University)

ADOP Advances in Optoelectronics

NONLINEAR DYNAMICS IN OPTICAL COMPLEX SYSTEMS

Kenju OTSUKA
Tokai University, Hiratsuka, Kanagawa, Japan

KTK Scientific Publishers / Tokyo

Kluwer Academic Publishers
Dordrecht / London / Boston

C.I.P. Catalogue record for this book is available from the Library of Congress.

ISBN 07923-6132-6 (Kluwer)

Published by KTK Scientific Publishers, 2002 Sansei Jiyugaoka Haimu, 27-19 Okusawa 5-chome, Setagaya-ku, Tokyo 158-0083, Japan / Kluwer Academic Publishers, P.O. Box 17, 3300 AA Dordrecht, The Netherlands.

Kluwer Academic Publishers incorporates
the publishing programmes of
D. Reidel, Martinus Nijhoff, Dr W. Junk and MTP Press.

Sold and Distributed in the U.S.A. and Canada
by Kluwer Academic Publishers,
101 Philip Drive, Assinippi Park, Norwell, MA 02061, U.S.A.
in Japan by KTK Scientific Publishers,
2002 Sansei Jiyugaoka Haimu, 27-19 Okusawa 5-chome, Setagaya-ku,
Tokyo 158-0083, Japan.

In all other countries, sold and distributed
by Kluwer Academic Publishers,
P.O. Box 322, 3300 AH Dordrecht, The Netherlands.

All Rights Reserved

© 1999 by KTK Scientific Publishers / Kluwer Academic Publishers

No part of the material protected by this copyright notice may be reproduced or utilized in any form or by any means, electronic or mechanical, including photo-copying, recording or by any information storage and retrieval system, without written permission from the copyright owner.

(This book is published by Grant-in-Aid publication of Scientific Research Result of the Ministry of Education, Science and Culture of Japan.)

Printed in Japan

PREFACE

In the beginning of this century, when there existed no computers in the world, Poincaré already imagined the existence of extremely complex trajectories of motions through his research on three-body problems and mentioned that *"trajectories are surprisingly complex and I do not intend to draw their pictures..."*. It took 60 years until unpredictability of such complex motions is recognized through research on weather [1] and motion of stars in the galaxy [2]. These two discoveries initiated the research field of nonlinear dynamics and chaos in science, and they are referred to as Lorenz chaos [1] and Henon-Heiles chaos [2]. Following these pioneering works, three universal routes to chaos, i.e., quasiperiodicity [3], period-doubling cascade [4] and intermittency [5], were discovered by early 80's. In accordance with these theoretical and mathematical progresses, nonlinear dynamics and chaos have been demonstrated experimentally in fields as diverse as chemistry, hydrodynamics, solid-state devices, biology, celestial mechanics, optics and so on. Rapid advancements in this field of nonlinear dynamics and chaos led to several discoveries of self-similarity laws and quantitive characterization methodology of chaotic motions (e.g., strange attractors). More recently, the research towards *coping with chaos* [6] utilizing the basic knowledge of the theory of chaos to achieve some practical goal, such as prediction, control and communication, has been demonstrated successfully.

Nonlinear dynamics and chaos in optics has a significant conceptual meaning, because fundamental models of nonlinear optical systems, which are derived from well-established Maxwell-Bloch equations, possess inherent instability leading to chaos. These models provide promising prototypes for investigating complex dynamical behaviors in strong connection with experimental demonstrations. The research on nonlinear optical dynamics forms the two poles together with the current topics in quantum optics such as squeezing, cavity quantum electrodynamics, laser cooling, Bose-Eistein condensation, quantum nondemolition measurement, so on and so forth, which present us a variety of microscopic quantum mechanical world.

Research on chaos in optics was triggered by the dicovery of the simple mathematical correspondence existing between Lorenz equations and Maxwell-Bloch laser equations with three variables by Haken in 1975 [7]. In early studies of solid-state lasers, period-doubling and chaotic spiking were already observed in a deeply modulated Nd:YAG laser and chaotic spiking oscillation was numerically observed in 1970 [8], although there was no concept of *chaos* in the field of Quantum Electronics. The proposal of Ikeda map in nonlinear passive optical resonators in 1979 [9], which exhibits a Feigenbaum's period-doubling bifurcation [4], observations of chaos in widely used practical laser diodes [10] and much more have accelerated the research of chaos in optics.

Generally speaking, the understanding of fundamental properties of chaos in small degrees of freedom systems such as Lorenz chaos has reached the period of maturity and interests are considered to shift towards dynamics of nonlinear systems with large degrees of freedom. Such systems are sometimes referred to as "complex systems" recently. The research along this line would give birth to new concepts and methodology in the next century in the process of systematic studies of complex systems which are far from the traditional condensed-matter physics which focuses on individual materials, mesoscopic and microscopic systems. Indeed, through our research of complex systems on the stage of quantum optics in recent years, the following generic properties have been recognized:

1. Complex systems are self-organized to preserve orders resulting from their nonlinearity, e.g., *vanishing gain circulation rule, antiphase dynamics, winner-takes-all dynamics and antiphase periodic states, universal power spectra relation, nonstationary chaos, etc.*

2. Complex systems self-create a variety of dynamic patterns, e.g., *chaotic itinerancy, self-formation of easy switching paths, majority-ruling switching, spot dancing, mode hopping, grouping chaos and cooperative synchronization, etc.*

3. Complex systems aquire various cooperative functions which are qualitatively different from individual elements, e.g., *domino dynamics, spatial chaos and felexible memory, factorial dynamic pattern memory, controlled switching-path formation and factorial pattern generation, parametric "linear" response, etc.*

This Monograph entitled "Nonlinear Dynamics in Optical Complex Systems," summarizes systematically our work on nonlinear dynamics and chaos in optics which has been done in these past 10 years at NTT Basic Research Laboratories, Université Libre de Bruxelles and Tokai University, focussing on nonlinear dynamics and cooperative functions in collective optical systems with large degrees of freedom. For this purpose, detailed derivations of traditional equations are spared to a certain extent and only key messages are included. The author would be more than happy if the readers could be stimulated by some generic nature in complex systems mentioned above.

The author is indebted to Profs. T. Kamiya and M. Ohtsu for providing him an opportunity of publishing this Monograph. He also thanks Prof. K. Ikeda of Ritsumeikan University, Prof. P. Mandel of Université Libre de Bruxelles (Belgium) and Prof. J.-L. Chern of National Cheng Kung University (Taiwan). Most of this book is written on the basis of articles co-authored by these collaborators. In particular, Sections 1.2 and 1.3 have been completed referring to discussions with Prof. Ikeda.

References

1) E. Lorenz, "Deterministic non-periodic flow", *J. Atmos. Sci.*, Vol. 20 (1963) pp. 130–141.

2) M. Henon and C. Heiles, "The applicability of the third integral of motion: some numerical examples", *Astron. J.*, Vol. 69 (1964) pp. 73–79.

3) D. Rulle and F. Takens, "On the nature of turbulence", *Commun. Math. Phys.*, Vol. 20 (1979) pp. 167–192.

4) M. J. Feigenbaum, "The onset spectrum of turbulence", *Phys. Lett.*, Vol. 74A (1979) pp. 375–378.

5) Y. Pomeau and P. Manneville, "Intermittent transition to turbulence in dissipative dynamical systems", *Comm. Math. Phys.*, Vol. 74 (1980) pp. 189–197.

6) *Coping with Chaos*, E. Ott, T. Sauer and J. A. York Eds., John Wiley & Sons, Inc. 1994.

7) H. Haken, "Analogy between higher instabilities in fluids and lasers", *Phys. Lett.*, Vol. A53 (1995) pp. 77–78.

8) T. Kimura and K. Otsuka, "Response of a cw Nd^{3+}:YAG laser to sinusoidal cavity perturbations", *IEEE J. Quantum Electron.*, Vol. QE-6 (1970) pp. 764–769.

9) K. Ikeda, "Multiple-valued stationary state and its instability of the transmitted light by a ring cavity system", *Opt. Commun.*, Vol. 30 (1979) pp. 257–261.

10) T. Mukai and K. Otsuka, "New route to optical chaos: successive-subharmonic-oscillation cascade in a semiconductor laser coupled to an external cavity", *Phys. Rev. Lett.*, Vol. 55 (1985) pp. 1711–1714.

Contents

Preface

1 PROLOGUE TO NONLINEAR DYNAMICS IN OPTICAL COMPLEX SYSTEMS 1

- 1.1 Fundamental Semiclassical Equations 2
- 1.2 Homogeneously-Broadened Single-Mode Laser 3
    - 1.2.1 Stationary states and linear stability analysis 3
    - 1.2.2 Lorenz-Haken instability 6
    - 1.2.3 Experimental evidence of laser Lorenz chaos 11
- 1.3 Inhomogeneously Broadened Single-Mode Laser 13
    - 1.3.1 Casperson instability 13
    - 1.3.2 Observation of three universal routes to chaos in a detuned system 20
- 1.4 Nonlinear Dynamics and Chaos in Semiconductor Lasers 22
    - 1.4.1 Inherent $\chi^{(3)}$ nonlinearity in laser diodes (LDs) 23
    - 1.4.2 Modulated extended cavity laser diodes 24
    - 1.4.3 Autonomous instabilities in composite cavity laser diodes 28
- 1.5 Period-Doubling Bifurcation in Nonlinear Passive Resonator 37
    - 1.5.1 Delay-differential equation model..................... 37
    - 1.5.2 Ikeda instability 39
    - 1.5.3 Physical interpretation of Ikeda instability 39
    - 1.5.4 Experimental observation of Ikeda instability 42
    - 1.5.5 Bifurcation and dynamical memory in the Ikeda-map systems... 42
- References... 49

2 REFERENCE MODELS OF LASER COMPLEX SYSTEMS — 53

2.1 Maxwell-Bloch Turbulence in Homogeneously-Broadened Multimode Laser Model — 54
    2.1.1 Resonant Rabi instability — 54
    2.1.2 Self-induced mode partition instability — 58
    2.1.3 Self-induced mode hopping: chaotic itinerancy — 63
    2.1.4 Physical interpretation and information theoretic analysis — 65

2.2 Modulation Dynamics of Globally Coupled Multimode Lasers — 66
    2.2.1 Introduction — 67
    2.2.2 Equations of motion and antiphase periodic states — 68
    2.2.3 Seeding-induced excitation of APS's — 71
    2.2.4 Grouping phenomenon in a large number of modes — 76
    2.2.5 Factorial dynamic pattern memory — 84

2.3 Chaotic Itinerancy in Antiphase Intracavity Second-Harmonic Generation (ISHG) — 88
    2.3.1 Introduction — 88
    2.3.2 Multimode ISHG equations and Hopf bifurcation — 88
    2.3.3 Bifurcation process and circulation analysis — 91

2.4 Clustering, Grouping, Self-Induced Switching and Controlled Dynamic Pattern Generation in an Antiphase Intracavity Second-Harmonic Generation Laser — 99
    2.4.1 Introduction — 99
    2.4.2 Q-switching antiphase periodic states in ISHG — 99
    2.4.3 Dynamical states associated with APS — 102
    2.4.4 Perturbation-induced switching-path formation and controlled dynamic pattern generation — 112

References — 117

3 ANTIPHASE DYNAMICS IN MULTIMODE LASERS — 119

3.1 Introduction — 119
3.2 Self-Organized Relaxation Oscillations — 120
    3.2.1 Multimode laser equations and linear stability analysis — 120
    3.2.2 Noise power spectra — 124
3.3 Antiphase Dynamics in Modulated Multimode Lasers — 127
    3.3.1 Modulation experiments — 128
    3.3.2 Numerical analysis — 131
    3.3.3 Vanishing gain circulation — 133

	3.4	Parametric Resonance in a Modulated Microchip Multimode Laser..	135
		3.4.1 Multichannel laser Doppler velocimetry modulation scheme..	136
		3.4.2 Periodic oscillations with multiple-parametric resonances...	138
		3.4.3 Numerical verifications................................	142
	3.5	Transverse Effects on Antiphase Dynamics.................	144
		3.5.1 Transverse effect on noise power spectra.............	145
		3.5.1 Modulation dynamics...............................	148
		References..	155

4 MORE ON MULTIMODE LASER DYNAMICS — 157

	4.1	Variation of Lyapunov Exponents on a Strange Attractor for Spiking Laser Oscillations.....................................	157
		4.1.1 Introduction..	157
		4.1.2 Local Lyapunov exponents and Allan variance........	158
		4.1.3 Nonstationary chaos in spiking multimode lasers.....	162
		4.1.4 Summary and future problems......................	168
	4.2	Suppression of Chaotic Spiking Oscillations.................	169
		4.2.1 Experimental results in a $LiNdP_4O_{12}$ microchip laser	169
		4.2.2 Numerical verification of suppressing chaos in a modulated multimode laser.......................	173
		4.2.3 Summary and outlook..............................	174
	4.3	Chaotic Burst Generation in a Compound Cavity Multimode Laser..	175
		4.3.1 Frustration phenomenon: periodic and chaotic intensity drop......................................	176
		4.3.2 Mode-partition-noise-induced chaotic bursts..........	181
		4.3.3 Summary and outlook..............................	185
	4.4	Breakup of CW Multimode Oscillations by High-Density Pumping..	185
		4.4.1 Introduction..	185
		4.4.2 Experimental results................................	186
		4.4.3 Theoretical results..................................	191
		4.4.4 Antiphase selfpulsations in the Λ-scheme.............	195
		References..	198

5 LASER ARRAY DYNAMICS — 201

	5.1	Class-A Laser Array Dynamics............................	201

	5.1.1	Optical analogue of discrete time-dependent complex Ginzburg-Landau equations	202
	5.1.2	$N=3$ low dimensional chaos: Complex Ginzburg-Landau attractor	203
	5.1.3	Self-induced path formation among local attractors	208
	5.1.4	Intermittent phase turbulence in continuum limit	214
	5.1.5	Switching paths in CTDGLs	216
	5.1.6	Chaotic itinerancy in open CTDGL systems	217
	5.1.7	Summary and discussion	218
5.2	Evanescent-Field Coupled Class-B Laser Array Dynamics		218
	5.2.1	Self-induced phase turbulence and chaotic itinerancy	218
	5.2.2	Stationary states	219
	5.2.3	Self-induced phase turbulence and spot dancing	221
	5.2.4	Self-induced chaotic relaxation oscillations and spot dancing in laser-diode arrays	224
	5.2.5	Chaotic itinerancy and super-slow relaxation	226
5.3	Globally-Coupled Class-B Laser Array Dynamics		230
	5.3.1	Factorial dynamical pattern memory in short-delay	231
	5.3.2	Delay-induced generalized bistability in one-element system	233
	5.3.3	Delay-induced instability in GCLA	240
	References		246

6 COOPERATIVE DYNAMICS AND FUNCTIONS IN COLLECTIVE NONLINEAR OPTICAL ELEMENT SYSTEMS   249

6.1	Nonlinear Polarization Dynamics and Spatial Chaos		250
	6.1.1	Polarization dynamics in crystal optics	250
	6.1.2	Spatial chaos in polarization	255
6.2	Spatiotemporal Dynamics of Coupled Nonlinear Optical Element Systems		261
	6.2.1	Otsuka-Ikeda model	262
	6.2.2	Firth model	275
6.3	Cooperative Functions of Coupled Nonlinear Element Systems		276
	6.3.1	All-optical signal processing in bistable chain	277
	6.3.2	Spatial chaos memory	281
	References		290

Index   291

Chapter 1

PROLOGUE TO NONLINEAR DYNAMICS IN OPTICAL COMPLEX SYSTEMS

In this chapter, fundamental model equations which describe the light-matter interaction are summarized. Then, several key optical systems, which initiated the research on nonlinear dynamics and chaos in optical systems, and their essential properties are reviewed as the basis for "optical complex systems" described in the following chapters.

Before discussing a variety of optical instabilities, it should be pointed out that there exist three characteristic frequencies which appear when we consider dynamics resulting from the light-matter interaction. It is interesting to note that instabilities associated with these frequencies appear under appropriate conditions.

These three frequencies are: (1) the relaxation oscillation frequency, (2) the longitudinal mode spacing frequency and (3) the Rabi precession frequency. Relaxation oscillations refer to oscillations resulting from the interplay between the photon number and the population inversion in lasers, such as solid-state lasers, CO_2 laser, and semiconductor lasers, in which the transverse relaxation rate is extremely large as compared with the longitudinal relaxation rate, i.e., $\gamma_\perp \gg \gamma_\|$. The longitudinal cavity mode spacing frequency corresponds to the frequency difference between longitudinal modes which resonate in the cavity. This frequency plays an important role for instabilities in the passive nonlinear resonator and corresponds to the fundamental frequencies in multimode oscillations as well. Finally, Rabi precession refers to coherent oscillations of material fields (i.e., polarizations) which take place under the applied electric field in far-infrared lasers.

A schematic diagram of time scales of optical instabilities under typical oscillation conditions is depicted in Fig. 1.1, showing the relation between characteristic frequencies and three relaxation rates, κ, $\gamma_\perp$ and $\gamma_\|$, which characterize the optical resonator and the nonlinear medium. Here, κ is

CHAPTER 1. PROLOGUE TO NONLINEAR DYNAMICS

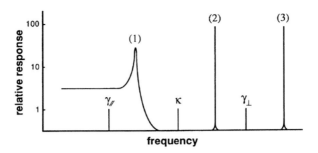

Figure 1.1: Schematic diagram of laser instabilities. Shown is the output modulation (peak intensity over average intensity) versus frequency of cavity loss modulation.

the damping rate of the optical cavity.

In section **1.1**, fundamental semiclassical equations which describe the light-matter interaction are described. Sections **1.2** and **1.3** describe laser instabilities and chaos associated with Rabi precession and the effect of inhomogeneous broadening on laser dynamics. Section **1.4** discusses instabilities and chaos related to relaxation oscillations. Finally, instabilities and chaos in passive nonlinear resonators are reviewed in section **1.5**.

1.1 Fundamental Semiclassical Equations

We consider atoms interacting with an "intense" electromagnetic field for which the quantization is not necessary. Starting from the Schrödinger equation, the following dynamical equations describing the light-matter interaction in two-level systems are derived in terms of density matrix elements ρ_{ij}:

$$d\rho_{12}/dt = (i\omega_0 - \gamma_\perp)\rho_{12} - i(\mu_{12}/h)E\Delta\rho, \qquad (1.1)$$

$$d\Delta\rho/dt = i(4\pi\mu_{12}/h)E(\rho_{12} - \rho_{12}^*) + \gamma_\parallel(\Delta\rho_0 - \Delta\rho), \qquad (1.2)$$

where $h\omega_0/2\pi = W_2 - W_1$ (W_i is the energy level, ω_0 is the atomic resonance frequency), μ_{12} is the electric dipole moment, $\Delta\rho = \rho_{22} - \rho_{11}$ and $\Delta\rho_0 = \rho_{22}^{(0)} - \rho_{11}^{(0)}$.

From Maxwell equations, on the other hand, wave equations for the electromagnetic field $E(\mathbf{r}, t)$ and material polarization $P(\mathbf{r}, t)$ in the cavity

$$E(\mathbf{r}, t) = \sum_n \tilde{E}_n(t) U_n(\mathbf{r}) e^{i\omega_n t} + c.c., \qquad (1.3)$$

$$P(\mathbf{r}, t) = \sum_n \tilde{P}_n(t) U_n(\mathbf{r}) e^{i\omega_n t} + c.c., \qquad (1.4)$$

are given by

$$dÉ_n/dt + [\kappa_n + i(\Omega_n - \omega_n)]\tilde{E}_n = i(\omega_n/2\epsilon_0)\tilde{P}_n, \tag{1.5}$$

assuming rotating-wave and slowly-varying envelope amplitude (SVEA) approximations. Here, κ_n is the damping rate, Ω_n is the cavity resonance frequency, ω_n is the oscillation frequency of the n-th mode and ϵ_0 is the dielectric constant.

The macroscopic material polarization and population inversion are expressed by $P = N\rho_{12}\mu_{12} + c.c.$ and $\Delta N = N\Delta\rho$ is the population inversion density, where N is the density of atoms. Assume a single-mode oscillation, then the following Maxwell-Bloch laser equations for the complex field amplitude $\tilde{E}$, the complex polarization amplitude $\tilde{\rho}$ oscillating at the frequency ω and $\Delta\rho$ are derived from Eqs. (1.1), (1.2) and (1.5)

$$d\tilde{E}/dt = -\kappa\tilde{E} - (N\mu\omega/2\epsilon_0)\tilde{\rho}, \tag{1.6}$$

$$d\tilde{\rho}/dt = [i(\omega - \omega_0) - \gamma_\perp]\tilde{\rho} - (2\pi\mu/h)\tilde{E}\tilde{\Delta}\rho, \tag{1.7}$$

$$d\tilde{\Delta}\rho/dt = \gamma_\parallel(\Delta\rho_0 - \tilde{\Delta}\rho) + (\pi\mu/h)(\tilde{E}\tilde{\rho}^* + \tilde{E}^*\tilde{\rho}), \tag{1.8}$$

where subscripts are omitted.

In the stationary state, from the relation $P = \epsilon_0\chi E$, the complex electric susceptibility χ is derived from the Bloch equations (1.6)–(1.8) as

$$\chi \equiv \chi' - i\chi'' = \frac{2\pi N\Delta\rho_0|\mu|^2[(\omega_0 - \omega) - i\gamma_\perp]}{\epsilon_0 h[(\omega_0 - \omega)^2 + \gamma_\perp^2 + (2\pi\mu E/h)^2(\gamma_\perp/\gamma_\parallel)]}. \tag{1.9}$$

Here, the real part expresses the dispersion (refractive index) profile and the imaginary part expresses the light amplification (absorption) profile. χ' and χ'' obey Kramers-Kronig relationship, where the term $(2\pi\mu E/h)^2(\gamma_\perp/\gamma_\parallel)$ is a light-intensity dependent $\chi^{(3)}$ nonlinearity of the refractive index and gain (absorption).

1.2 Homogeneously-Broadened Single-Mode Laser

1.2.1 Stationary states and linear stability analysis

Let us consider the case that the oscillation frequency ω is tuned to the atomic transition frequency ω_0, i.e., $\omega = \Omega = \omega_0$. Here, we introduce a new variable $F = -(2\pi\mu/h)\tilde{E}$ and notations are changed as $\tilde{\rho} \to \rho$ and $\tilde{\Delta}\rho \to w$. In this case, Eqs. (1.6)–(1.8) are written as

$$dF/dt = -\kappa F + s^2\rho, \tag{1.10}$$

$$d\rho/dt = -\gamma_\perp\rho + Fw, \tag{1.11}$$

$$dw/dt = \gamma_\|(w_0 - w) + F\rho, \tag{1.12}$$

where $s^2 \equiv N\omega|\mu|^2/2\epsilon_0$.

We abbreviate the set of 3 variables (F, ρ, w) of the Maxwell-Bloch equations (1.10)–(1.12) with vector $\mathbf{x}$ and let Eqs. (1.10)–(1.12) be written formally by

$$d\mathbf{x}/dt = \mathbf{F}(\mathbf{x}) \tag{1.13}$$

$\mathbf{F}(\mathbf{x})$ defines the flow vector at the point $\mathbf{x}=(F, \rho, w)$ in the 3-dimensional phase space whose components are the r.h.s. of Eqs. (1.10)–(1.12). Stationary solutions, i.e., fixed point $\mathbf{x}_s$ is the point which satisfies $d\mathbf{x}_s/dt = 0$, i.e.,

$$\mathbf{F}(\mathbf{x_s}) = \mathbf{O}$$

The problem is the stability of $\mathbf{x}_s$. The infinitesimal deviation $\delta \mathbf{x} = (\delta F, \delta \rho, \delta w)$ around $\mathbf{x}_s$ obeys the following linearized equation of motion

$$d\delta\mathbf{x}/dt = (\partial \mathbf{F}(\mathbf{x_s})/\partial \mathbf{x})\delta\mathbf{x} \tag{1.14}$$

$\partial \mathbf{F}(\mathbf{x_s})/\partial \mathbf{x}$ represent a 3×3 Jacobian matrix with (i,j) components given by $\partial F_i(\mathbf{x_s})/\partial \mathbf{x_j}$. The stability analysis of the model equations (1.10)–(1.12), is quite simple, and is left as an exercise for the readers. Only the final results are shown below:

(1) If the linear gain $\alpha_0 = s^2 w_0/\gamma_\perp$ is smaller than the threshold value $\alpha_{th}^{(1)} = \kappa$, only the trivial non-lasing solution,

$$\mathbf{O} : (F_s, \rho_s, w_s) = (0, 0, w_0) \tag{1.15}$$

exists as a stable fixed point solution.

(2) If α_0 exceeds the first theshold $\alpha_{th}^{(1)}$, $\mathbf{O}$ becomes unstable, and the lasing solution

$$\mathbf{L}^\pm : (F_s, \rho_s, w_s) = (\pm\sqrt{\gamma_\|\gamma_\perp(\frac{\alpha_0}{\sigma} - 1)}, \pm\frac{\kappa}{s^2}\sqrt{\gamma_\|\gamma_\perp(\frac{\alpha_0}{\sigma} - 1)}, \frac{\gamma_\perp \kappa}{s^2}) \tag{1.16}$$

appeas as a new stable fixed points. At $\alpha_0 = \alpha_{th}^{(1)}$, the stability exponents at $\mathbf{O}$ is

$$\lambda_1 = 0, \quad \lambda_2 = -\gamma_\| \quad \lambda_3 = -(\kappa + \gamma_\perp), \tag{1.17}$$

and λ_1 becomes positive at $\alpha_0 > \alpha_{th}^{(1)}$.

(3) If α_0 is increased further and under the condition

$$\kappa > \gamma_\| + \gamma_\perp, \tag{1.18}$$

1.2. HOMOGENEOUSLY-BROADENED SINGLE-MODE LASER

even the fixed points $\mathbf{L}^\pm$ become unstable when α_0 exceeds the "seccond" threshold. The ratio of $\alpha_{th}^{(2)}$ to $\alpha_{th}^{(1)}$ is given by

$$\alpha_{th}^{(2)}/\alpha_{th}^{(1)} = \frac{\kappa}{\gamma_\perp}(\frac{\kappa}{\gamma_\perp} + \frac{\gamma_\|}{\gamma_\perp} + 3)/(\frac{\kappa}{\gamma_\perp} - \frac{\gamma_\|}{\gamma_\perp} - 1). \tag{1.19}$$

and the stability exponents around $\mathbf{L}^\pm$ at $\alpha_0 = \alpha_{th}^{(2)}$ are given by,

$$\lambda_1 = -(\kappa + \gamma_\| + \gamma_\perp), \quad \lambda_{2,3} = \pm i\sqrt{2\kappa(\kappa + \kappa_\perp)(\kappa - \gamma_\| - \gamma_\perp)/\gamma_\perp} \tag{1.20}$$

For $\alpha_0 > \alpha_{th}^{(2)}$, the real part of $\lambda_{2,3}$ becomes positive, and a precession at frequency $Im(\lambda_{2,3})$ is excited. Consequently, the bifurcation phenomenon which occurs at $\alpha_0 = \alpha_{th}^{(2)}$ is a Hopf bifurcation.

The above descriptions are too much arithmetic. We therefore present a physical picture of the instability above the second threshold since the indicated phenomenon contains a typical aspect of laser instabilities.

For the sake of simplicity, let us consider the special case of $\gamma_\| = \gamma_\perp \equiv \gamma$, and introduce a complex variable Z representing the material field

$$Z = s^2(\rho - iw). \tag{1.21}$$

The linearized equation (1.14) is then written as

$$d\delta F/dt = -\kappa\delta F + (\delta Z + \delta Z^*)/2 \tag{1.22a}$$

$$d\delta Z/dt = -\gamma\delta Z + iZ_s\delta F + iF_s\delta Z \tag{1.22b}$$

Let us now consider the limit of large linear gain $\alpha_0 = s^2 w_0/\gamma_\perp$. The third term on the right side of equation (1.22b) becomes dominant. This means that the deviation δZ of the material field starts to oscillate as $\delta Z \sim \tilde{\delta Z} e^{iF_s t}$, where $\tilde{\delta Z}$ is the slowly varying part of δZ. This oscillation is the Rabi precession of the material field. The precession components of the material field, via the Maxwell equation (1.22a), induce the oscillation of electric field at the same frequency. Let $\delta F \sim \tilde{\delta F} e^{iF_s t}$, then $\tilde{\delta F} \sim \tilde{\delta Z}/2(iF_s + \kappa)$. In short,

$$\delta F \sim \frac{\delta Z}{2(iF_s + \kappa)}. \tag{1.23}$$

Note that the iF_s term resulting from the Rabi precession is included in the denominator. In the limit of $\alpha_0 \gg \kappa$, i.e., $F_s \to \infty$, this effect is significant and δF possesses only the out-of-phase component of δZ. In other words, induced electric fields are dispersive.

Therefore, the dispersive component of electric field induced by Rabi precession is fedback to the material field via the second term of equation (1.22b). This term yields

$$\frac{iZ_s}{2(iF_s + \kappa)}\delta Z \longrightarrow \frac{\kappa}{2}\delta Z.$$

This implies that the effective decay constant of the material field is renormalized to be $-\gamma_\perp + \kappa/2$. Hence, if the inequality

$$\kappa > 2\gamma \tag{1.24}$$

is satisfielld, the Rabi precession will oscillate. The condition (1.24) is none other than the condition (1.18). Consequently, what is occuring above the second threshold is a self-induced Rabi precession.

The condition (1.18) or (1.24) requires that the electric field decay time (κ^{-1}) of the resonator should be shorter than the material field decay time ($\gamma_\perp^{-1}, \gamma_\parallel^{-1}$) and it is often referred to as "bad cavity" condition. In most lasers, this condition is not satisfied. Furthermore, from equation (1.19), the value of the second threshold is much larger than that of the first threshold. Indeed, in the case of $\gamma_\perp = \gamma_\parallel$, we may evaluate $\alpha_{th}^{(2)}/\alpha_{th}^{(1)} \geq 8 + 2\sqrt{12} \sim 15$. In past experiments, laser instabilities have been observed at the similar pumping levels as the first threshold value. Moreover, there is no guarantee that the single mode laser model is valid even at such a high pumping level. From the points mentioned above, it seems unlikely that the instability of the homogeneously-broadened single-mode laser is realistic.

However, conceptually, this instability has an important meaning. One reason is that this is an instability phenomenon exhibited by the simplest model system which is recognized by all physicists to express fundamental laser dynamics. Another reason is that this instability results in a chaotic phenomenon which provides a prototype of complex dynamic phenomena described by a simple ordinary differential equation with only 3 variables. In the next subsection we will discuss the dynamical phenomenon that occurs above the second threshold.

1.2.2 Lorenz-Haken instability

A typical trajectory of motion which is finally realized under the conditions $\alpha_0 > \alpha_{th}^{(2)}$ is depicted in the $F\rho w$ - phase space in Fig. 1.2. Temporal variation of F is also shown. It is clear that two rotating motions exist surrounding symmetric centers with respect to $F = \rho = 0$ and switching occurs erratically between the two rotating motions, yielding an extremely

1.2. HOMOGENEOUSLY-BROADENED SINGLE-MODE LASER

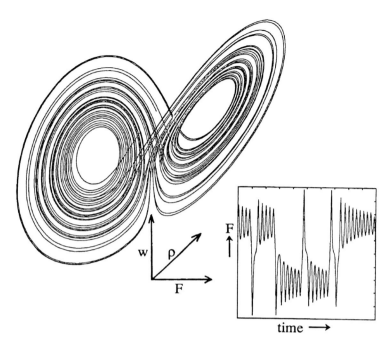

Figure 1.2: Trajectory of the motion on the Lorenz sttrange attractor described by Eqs. (1.10)–(1.12). The insert is temporal variation of $F(t)$.

complex motion. Both centers correspond to the lasing solutions $\mathbf{L}^{\pm}$. The rotating motion around these equilibria is the unstable Rabi precession discussed in the previous section.

The first person who discovered such a curious behavior shown by the solution of the laser rate equation was not a laser physicist. In 1963, a meterologist, Lorenz proposed a simple model equation which describes the motion of the earth's atmosphere by truncating both Navier-Stokes equation and thermal convection equation [1]. This equation, called the Lorenz Equation, has 3 variables x, y, z and is expressed by the following set of ordinary differential equations:

$$\dot{x} = k(y - x) \tag{1.25a}$$

$$\dot{y} = \alpha x - y - xz \tag{1.25b}$$

$$\dot{z} = -\beta z + xy \tag{1.25c}$$

13 years later, Haken realized that this equation is equivalent to the

laser rate equation [2]. Indeed, by the following simple transformation

$$x \mapsto F/\gamma_\perp \quad y \mapsto \frac{s^2}{\kappa\gamma_\perp}\rho \quad z \mapsto -\frac{s^2}{\kappa\gamma_\perp}(w-w_0), \qquad (1.26)$$

and the replacement of the three parameters k, α, β by

$$k \mapsto \kappa/\gamma_\perp \quad \alpha \mapsto \frac{w_0 s^2}{\kappa\gamma_\perp} \quad \beta \mapsto \frac{\gamma_\|}{\gamma_\perp} \qquad (1.27)$$

Eq. (1.25) completely coincides with the Maxwell-Bloch laser rate equations (1.10)–(1.12)! Lorenz discovered through numerical analysis that the solution of equation (1.25), (or equivalently (1.10)–(1.12)) exhibits irregular behavior above the second threshold value $\alpha_{th}^{(2)}$. Furthermore, he gave a strong numerical proof showing that this behavior is not due to numerical error but is due to the essential characteristic of the system.

Starting from arbitrary initial conditions except on the unstable manifolds, the orbit is attracted onto a manifold in the phase space. Let us consider the temporal change of the volume element $\delta V = \delta F \delta \rho \delta w$. δV changes with the rate $d\delta V/dt/\delta V$ and it is given by the trace $\sum_i \partial F_i(\mathbf{x})/\partial x_i$ of the Jacobian matrix. In our system expressed by (1.10)–(1.12), this quantity is constant everywhere in the phase space

$$\sum_i \partial F_i(\mathbf{x})/\partial x_i = -(\kappa + \gamma_\| + \gamma_\perp) \qquad (1.28)$$

Here, the volume decreases as $e^{-(\kappa+\gamma_\|+\gamma_\perp)t}$ everywhere in the phase space. Consequently, the attracting manifold cannot possess a three dimensional volume. In fact, it can be well approximated by the branched "sheet" S. Let us indicate the geometrical structure of S in Fig. 1.3.

S can be interpreted as composed of three parts, S^+, S^-, S_0, and the underlying mechanism which comprises S from the 3 parts can be well understood by considering the local behavior near the 3 unstable fixed points in our system. They are the non-lasing states $\mathbf{O}$ and the two lasing set of states $\mathbf{L}^\pm$. (Fig. 1.3) Note that the stability exponents $(\lambda_1, \lambda_2, \lambda_3)$ of $\mathbf{O}$ have signs $(+,-,-)$ at $\alpha > \alpha_{th}^{(2)}$. Writing the eigenvector corresponding to λ_i as $\mathbf{e}_i(\mathbf{O})$, we see that the trajectory first approaches $\mathbf{O}$ being attracted toward the plane spanned by $\mathbf{e}_2(\mathbf{O}), \mathbf{e}_3(\mathbf{O})$, and then is pushed out of $\mathbf{O}$ along the direction of $\mathbf{e}_1(\mathbf{O})$. Thus $\mathbf{O}$ is a hyperbolic fixed point. On the other hand, the real parts of the stability exponents $(\lambda_1, \lambda_2, \lambda_3)$ have signs $(-,+,+)$ at $\mathbf{L}^\pm$. Furthermore, since λ_2 and $\lambda_3 (= \lambda_2^*)$ have an imaginary part, the orbit approaches $\mathbf{L}^\pm$ from the direction of $\mathbf{e}_1(\mathbf{L}^\pm)$ and is pushed out of $\mathbf{L}^\pm$, exhibiting the precession on the plane spanned by $\mathbf{e}_2(\mathbf{L}^\pm)$ and $\mathbf{e}_3(\mathbf{L}^\pm)$.

1.2. HOMOGENEOUSLY-BROADENED SINGLE-MODE LASER

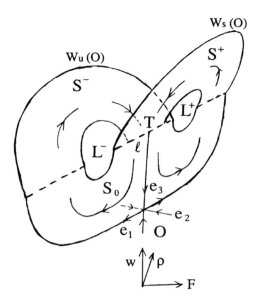

Figure 1.3: Branched sheet S modelling the attracting manifold (Lorenz attractor).

As shown in Fig. 1.3, let us start from the vicinity of $\mathbf{O}$ along $\mathbf{e_1(O)}$ (or $-\mathbf{e_1(O)}$). This orbit defines the unstable manifold $W_u(\mathbf{O})$ of $\mathbf{O}$. When $W_u(\mathbf{O})$ is separated from $\mathbf{O}$, it enters the attracting domain of $\mathbf{L}^+$ (or $\mathbf{L}^-$), and approaches $\mathbf{L}^+$ (or $\mathbf{L}^-$) from the direction $\mathbf{e_1(L^+)}$ (or $\mathbf{e_1(L^-)}$). Next, within the plane made up of $\mathbf{e_{2,3}(L^\pm)}$, precession continues, and then the orbit is pushed out. In this manner, $W_u(\mathbf{O})$ forms the boundary of S. Roughly speaking, $S^\pm, S_0$ can be interpreted as the manifolds obtained by extending the planes spanned by $\mathbf{e_{2,3}(L^\pm)}$ and $\mathbf{e_{1,3}(O)}$, respectively. S^+, S^- and S_0 meet at the line ℓ. In other words, the orbit precessing on S^+ and the orbit precessing on S^- must meet at ℓ. This seems to contradict the one-to-one correspondence between the intial state and the final state required from the ordinary differential equation. This contradiction implies that S_0 (and so $S^\pm$) is not a single sheet, but it must have multifolded structures. In fact, S_0 has a transversal structure, in which the thickness is too thin to be resolved. If this thickness is ingnored, S can be approximated by a single sheet. Such a physical picture is established as far as the forward time evolution is concerned.

Consider the sheet $\mathbf{W}_s(\mathbf{O})$ which is formed by extending the stable manifold of $\mathbf{O}$, i.e., the manifold originating from $\mathbf{e_2}$ and $\mathbf{e_3}$ in accordance with the time reversal process. $\mathbf{W}_s(\mathbf{O})$ intersects ℓ. Let the intersecting point be T. As shown in Fig. 1.4(a), let us start from a point on S^-.

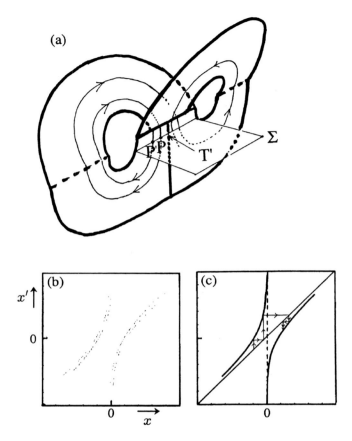

Figure 1.4: (a) Motion on the branched sheet S, (b) numerically constructed return map, and (c) approximate one-dimensional return map. In (b) and (c), x is defined as $x = +PT'$ (righthand side of T') or $-PT'$ (lefthand side).

The orbit precesses around $\mathbf{L}^-$ inside the sheet formed by S^- and S_0. After one revolution, it intersects with ℓ without a failure. The "fate" of the remaining orbit is determined by whether the intersecting point is on the left or the right of T. If it is on the left, revolutions inside the $S^- + S_0$ sheet are once again continued. However, since the revolution radius increases, the orbit will have to intersect with ℓ on the right side. Consequently, the orbit is entered into the complementary sheet $S^+ + S_0$, and is switched to precessions inside this sheet. In this way, due to the combination of the two precessions (i.e., Rabi nutation) inside two branched sheets, i.e., $S^+ + S_0$ and $S^- + S_0$, and the switching mechanism at T, the chaotic motion, featuring erratic switching of the precession around two

1.2. HOMOGENEOUSLY-BROADENED SINGLE-MODE LASER

symmetrical rotation centers ($\mathbf{L^+}, \mathbf{L^-}$), is realized.

To give an understanding of the properties of chaotic motion mathematically, let us form a return map. Consider a two-dimensional surface Σ which traverses S at ℓ or in its vicinity as illustrated in Fig. 1.4(a). Let the point where the orbit crosses Σ be P and suppose the point P' where the orbit once again crosses Σ at the next time. The mapping $P \mapsto P'$ defines a 2-dimensional map, called a Poincaré map. However in our system, as the orbit is approximately on a single sheet S, this map is reduced to a one-dimensional map between points on ℓ. With T' as the origin, let the position of P and P' be x, x', respectively. In Fig. 1.4(b) we show an example of numerically constructed return map. It is evident that it has (infinitely) multiple structure, but its thickness is so thin that we may approximate the return map by a single valued function $x' = f(x)$ (Fig. 1.4(c)). The relationship between the return map and the orbit on sheet S is easily understood from Fig. 1.4(a) and (b). The return map reflects the symmetry of the system and consists of two symmetric branches with respect to the origin.

The $x < 0$ ($x > 0$) branch shows Rabi precession around $\mathbf{L^-}$ ($\mathbf{L^+}$). The derivative $|f'(x)|$ of each of the branches is always greater than 1, since the orbital radius of precession is expanding. No matter how close the two intial values x_0, x_0' are, the distance of the n-th iteration $f^n(x_0)$ and $f^n(x_0')$ expands exponentially, and at a certain n, $f^n(x_0)$ and $f^n(x_0')$ will belong to two different branches; and eventually $f^n(x_0)$ and $f^n(x_0')$ lose their correlation. The above properties are the essential nature of Lorenz chaos.

If α_0 is far above the second threshold $\alpha_{th}^{(2)}$, the chaotic oscillation ceases and a pair of asymmetric periodic solutions finally appears through an inverse process of the period doubling. However, the bifurcation scenario is not restricted to this one and there is much more; homoclinic explosion for instance.

1.2.3 Experimental evidence of laser Lorenz chaos

In most laser systems, the transverse relaxation rate $\gamma_\perp$ is much larger than the resonator linewidth κ. Consequently, it is not common that the fundamental condition (1.18) or (1.24) for the second threshold is satisfied. However, in lasers which oscillate in far-infrared regions, the spontaneous relaxation rate which determines the lower limit of $\gamma_\perp$ can be very small because the spontaneous lifetime is proportional to (oscillation frequency)$^{-3}$. Far-infrared lasers also have a large linear gain. This fact makes it possible for the second threshold to appear in far-infrared (FIR) lasers in general.

The possibility of observing Lorenz-like chaos was suggested by Zeghlache et al. [3] and the occurrence of Lorenz-like chaos was experimentally

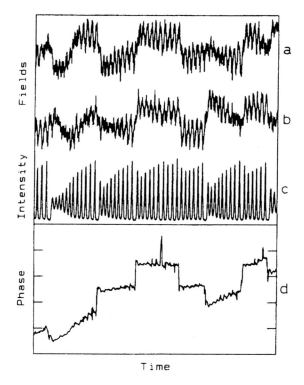

Figure 1.5: High pressure ($p = 9$ Pa) chaotic pulsing of 153-μm ^{15}NH$_3$ laser emission for resonant tuning. Trace marked "Intensity": laser pulse intensity. Pulsing period is 1 μs. Traces marked "fields": in-phase and in-quadrature heterodyne signals measuring the laser field. Trace marked "phase": phase changes of the laser field as a function of time, reconstructed from field traces. One division on the vertical axis corresponds to phase change of π rad.

confirmed to exist under the appropriate pressure and pump range [4]. The rotational levels of molecules such as CO$_2$, NH$_3$ etc. are often utilized for the FIR laser operation. Excitation to the lasing levels is usually done by coherent laser source. However, there is a possibility that such a coherent pumping connects the ground state with the two lasing levels, forming a three level system. On the other hand, the basic Lorenz system models an incoherently pumped two level laser and the pump induced coherence between levels may possibly violate the validity of the two-level description. According to an analysis of a three level model, the transition to chaos takes place gradually via successive bifurcation of periodic states and the abrupt transition to Lorenz chaos predicted by the laser rate equation can not be seen. Furthermore, the shape of the chaotic attractor is asymmetric and the symmetric attractor indicated in Fig. 1.2 does not appear. However, under high pressure, collision between the molecules destroys the pump induced coherence, with the two level descriptions being restored once again, and the possibility of observing a symmetric Lorenz attractor as shown in Fig. 1.2 arises.

Weiss *et al.* used the transition between the rotational levels of NH_3 molecule for a FIR laser, and they discovered that along with an increase in pressure, Lorenz-like chaos could certainly be observed [4]. Temporal evolutions of the observed electric filed amplitude and electric field phase are shown in Fig. 1.5. It is clear that in the high pressure regime, the phase of electric field changes abruptly by $\pm\pi$. In other words, random switching of the sign of electric field, which is a remarkable characteristic of Lorenz chaos, was observed. In a low pressure reigme, this kind of behavior cannot be observed [4]. Instead, asymmetric attraction predicted by a three level single mode laser model including the pumping coherence effect was observed. This implies that in a low pressure regime, molecular collision does not suppress coherence of a three level system.

In addition, when the oscillation line detunes from the transition frequency, in other words, in the off-resonance case ($\omega \neq \omega_0$), the symmetric attractor is destroyed again. In this case, occurrence of period-doubling predicted by a numerical simulation [5] has been experimentally confirmed with the NH_3-FIR laser [6].

1.3 Inhomogeneously Broadened Single-Mode Laser

1.3.1 Casperson instability

The instability threshold of homogeneously broadened lasers is, in general, much larger than the threshold of laser oscillation. On the other hand, the occurrence of an instability near the laser oscillation threshold was pointed out by Casperson and his co-workers for He-Xe gas lasers. He intensively studied this peculiar instability and succeeded in elucidating the mechanisms experimentally and theoretically [7]. This subsection will discuss the Casperson instability.

In gas lasers, the atoms contributing to laser action move thermally and the transition line is inhomogenously broadened due to the Doppler effect. In other terms, each atom has a different resonant frequency. Since the atom velocity v obeys the law, $exp(-mv^2/2k_BT)$ (m: atom mass, k_B: Boltzman constant, T: temperature), localized macroscopic polarization is regarded as the summation from atom groups with the frequency distribution and the gain profile is given by

$$g(\omega) \propto \exp-[(\omega - \omega_0)/\Delta\omega_i)^2 ln2]. \tag{1.29}$$

Here, $\Delta\omega_i = kv^* = k\sqrt{2k_BT/m}$ is the inhomogeneous linewidth, where k is the wavenumber and v^* is the thermal velocity.

This effect induces a new type of instability different from the Lorenz instability. To be more specific, mathematically, both instabilities are born from Hopf bifurcations. The difference concerns the nature of the solution near and far away from the Hopf bifurcation point. The physical parameters which control the Hopf bifurcations are different. The analysis of the instability for the case of inhomogeneous broadening can, in principle, be achieved through the standard linear stability analysis but it is rather tedious. We wish to understand the physical origin of this instability without going into the details of stability analysis. Let us explain its essence in terms of laser spectroscopy.

Let us consider an inhomogeneously broadened laser system oscillating stationarily. For the sake of simplicity, we consider a single mode laser system whose resonance frequency Ω is tuned to the transition center frequency ω_0. In this case, the motion of the system can be expressed by the equations (1.6)–(1.8), setting $\omega_0 = \Omega$. Here, we introduce F again similarly to Eqs. (1.10)–(1.12). Assume that a sufficiently weak component δF which oscillates at frequency ν is produced in the electric field. Then, this oscillating component, through the Bloch equation (1.7)–(1.8), induces a weak component $\delta\rho$ in the material field which oscillates at the same frequency. The coupling between F and ρ by a beat note induces further the population pulsation component δw which oscillates at freqency ν in the population inversion. We write such behaviors by

$$\begin{pmatrix} \delta F \\ \delta \rho \\ \delta w \end{pmatrix} = e^{i\nu t} \begin{pmatrix} \tilde{\delta F}_+ \\ \tilde{\delta \rho}_+ \\ \tilde{\delta w}_+ \end{pmatrix} + \bar{e}^{i\nu t} \begin{pmatrix} \tilde{\delta F}_- \\ \tilde{\delta \rho}_- \\ \tilde{\delta w}_- \end{pmatrix}. \quad (1.30)$$

$\tilde{}$ was added to express the d.c. component. If the motion of $\tilde{\delta\rho}_\pm, \tilde{\delta w}_\pm$ is slow, the following equations are obtained from the linearized version of Bloch equation (1.7)–(1.8).

$$\tilde{\delta\rho}_- = D_\perp(\omega - \Omega - \nu)(F_s\tilde{\delta w}_- + w_s\tilde{\delta F}_-), \quad (1.31)$$

$$\tilde{\delta w}_- = D_\parallel(\omega - \Omega - \nu)(F_s\tilde{\delta\rho}^*_+ + F_s^*\tilde{\delta\rho}_- + \rho_s^*\tilde{\delta F}_- + \rho_s\tilde{\delta F}^*_+), \quad (1.32)$$

where $D_{\perp,\parallel}(\nu) = \frac{1}{i\nu+\gamma_{\perp,\parallel}}$.

To explain the essence of Caperson's instability, let us neglect the population pulsation contribution $\tilde{\delta w}_-$ for a while. Then, the Maxwell equation (1.6) can be written in the following form:

$$d\tilde{\delta F}_-/dt = [i\nu - \kappa + \tilde{\alpha}(\nu) + i\tilde{\beta}(\nu)]\tilde{\delta F}_-, \quad (1.33)$$

1.3. INHOMOGENEOUSLY BROADENED SINGLE-MODE LASER

where

$$\tilde{\alpha}(\nu) = s^2 \int_{-\infty}^{\infty} w_s(\omega) g(\omega) Re D_\perp(\omega - \Omega - \nu) d\omega, \quad (1.34)$$

$$\tilde{\beta}(\nu) = s^2 \int_{-\infty}^{\infty} w_s(\omega) g(\omega) Im D_\perp(\omega - \Omega - \nu) d\omega, \quad (1.35)$$

$\tilde{\alpha}(\nu)$ and $\tilde{\beta}(\nu)$ expresses the amplification coefficient and wavenumber shift, i.e. dispersion, which is produced due to the existence of the laser medium.

Therefore it can be easily understood that the condition for the oscillation at frequency ν to grow is given by

$$\tilde{\alpha}(\nu) > \kappa. \quad (1.36)$$

According to the definition, $\delta\tilde{F}$ does not have an oscillatory component. From this condition, i.e.,

$$\tilde{\beta}(\nu) = -\nu, \quad (1.37)$$

the oscillation frequency ν is determined. This condition physically implies that the wavenumber shift ν/c cancels the wavenumber shift $\beta(\nu)/c$ resulting from dispersion of the laser medium and a cavity resonance is achieved accordingly.

The net distribution function of population inversion, which are in the r.h.s. of Eqs. (1.34)–(1.35), i.e.,

$$\tilde{g}(\omega) \equiv g(\omega) w_s(\omega),$$
$$= g(\omega) \frac{(\Omega - \omega)^2 + \gamma_\perp^2}{(\Omega - \omega)^2 + \gamma_\perp^2 + \gamma_\perp |F_s|^2/\gamma_\parallel}, \quad (1.38)$$

has a "hole" near the atomic center frequency $\Omega = \omega_0$, as shown in Fig. 1.6(ai). This "hole" is burned because the lasing mode stimulates the resonating atoms to make transition to the lower level. This phenomenon is called spectral hole-burning. Due to spectral hole-burning, the gain $\tilde{\alpha}(\nu)$ peaks on both sides of the central frequency Ω. However, on the other hand, the existence of resonant atoms causes an anomaly in the phase shift $\tilde{\beta}(\nu)$ because of the dispersion effect, as illustrated in Fig. 1.6(ai). Consequently, frequencies $\nu_{\pm R}$, which satisfy $\tilde{\beta}(\nu_{\pm R}) = -\nu_{\pm R}$, appear on both sides of the central frequency Ω because of the spectral hole-burning as shown in Fig. 1.6(aii). If the gain at $\nu_{\pm R}$ is large enough, the $\nu = \nu_{\pm R}$ frequency components begin to oscillate. This *mode-splitting* phenomenon is the basic mechanism for Casperson instability.

In this manner, the spectral hole-burning and the anomalous dispersion bring about instability in inhomogeneously broadened lasers. However, in homogeneously broadened lasers, spectral hole-burning does not occur, as

16   CHAPTER 1. PROLOGUE TO NONLINEAR DYNAMICS

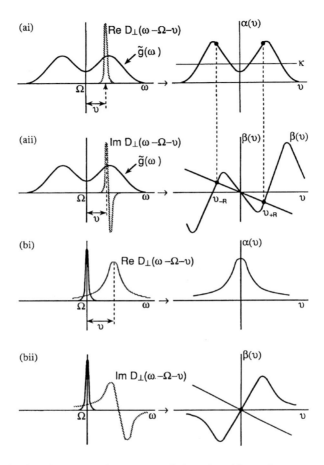

Figure 1.6: Mode splitting in inhomogeneously broadened lasers in comparson with homogeneously broadened lasers; (a) inhomogeneous broadening (b) homogeneous broadening.

shown in Fig. 1.6(b). Furthermore, the sign of anomalous dispersion is reversed, and the solution satisfying the resonance condition (1.37) does not exist except for the trivial solution of $\nu = 0$ (see Fig. 1.6(bii)). Lorenz instability in homogeneously broadened lasers is a higher order effect caused by the population pulsation δw, which was ingnored in the above discussion. As was discussed in section **1.3**, if we recall that Lorenz instability results from the Rabi precession which is the simultaneous oscillation of polarization and population inversion, it is no surprise that ignoring the population pulsation effect will not bring Lorenz instability. Even in the

1.3. INHOMOGENEOUSLY BROADENED SINGLE-MODE LASER

Caperson instability, the population pulsation effect has a significant effect on the instability threshold as indicated by Hendow and Sargent [8].

Next, the effect of population pulsation is examined. Let us substitute equations (1.31)–(1.32) into the Maxwell equation. Then, formally, the same equations as (1.34)–(1.35) are obtained, where the complex amplification rate is modified as follows:

$$\tilde{\alpha}(\nu) + i\tilde{\beta}(\nu)$$
$$= s^2 \int_{-\infty}^{\infty} g(\omega) D_\perp(\omega - \Omega - \nu)[w_s(\omega) + F_s \frac{\tilde{\delta w}_-(\omega)}{\tilde{\delta F}_-(\omega)}] d\omega, \quad (1.39)$$

The newly added second term in [ ] expresses the population pulsation effect. The ratio $\tilde{\delta w}_-/\tilde{\delta F}_-$ can be immediately determined from Eqs. (1.34)–(1.35) under the condition that $\tilde{\delta F}_+^* = \tilde{\delta F}_-$, i.e., δF_- is real. The instability threshold obtained through this method is much lower than when population pulsation is neglected. Note that these procedures lead to the same results as the linear stability analysis.

The numerically calculated dependence of the second threshold value

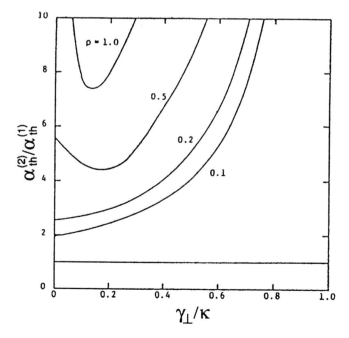

Figure 1.7: Dependence of the second threshold $\alpha_{th}^{(2)}$ on $\gamma_\perp/\kappa$ for different $\rho = \gamma_\parallel/\gamma_\perp$. $\gamma^*/\gamma_\perp = 1.0$.

on lasing parameters is shown in Fig. 1.7. In the limit of good cavity condition, the instability threshold becomes infinite, as in homoegeneously broadened systems. However under the bad cavity condition, the larger γ^* becomes, the nearer the threshold approaches the first threshold, where γ^* is the relaxation rate which is proportional to $\Delta\omega_i^{-1}$. According to Mandel, in the case of $\gamma^*/\gamma (\equiv \gamma_\perp = \gamma_\parallel) \gg 1$, the following relation is asymptotically obtained in the bad cavity limit [9]:

$$\alpha_{th}^{(2)} = (1 + \frac{2}{3}\frac{\gamma}{k})\alpha_{th}^{(1)} \qquad (1.40)$$

The transition process to chaos after the Casperson instability is conspicuously different from the Lorenz chaos. First of all, the oscillation at freqency ν is realized and the transistion to chaos seems to take place gradually. This process was experimentally and theoretically investigated in detail by Casperson. A He-Xe laser was used in the experiments and numerical calculations were carried out based upon a theoretical model,

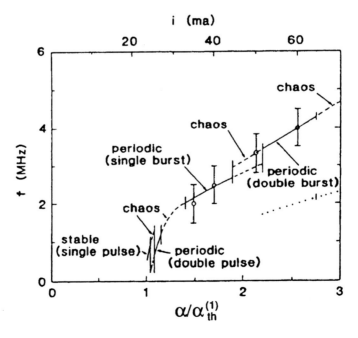

Figure 1.8: Theoretical (lines) and experimental (circles) pulsation frequencies as a function of pumping power. The dashed lines indicate the regions of chaotic output, and the dotted line is intended to emphasize the fact that in the period-doubled region the true repetition frequency is one half of what is indicated by the solid line.

1.3. INHOMOGENEOUSLY BROADENED SINGLE-MODE LASER

in which the effect of velocity changing collision is introduced into Eqs. (1.6)–(1.8). First, let us show the theoretical results. Figures 1.8 and 1.9 show the changes of the pulsation frequency and the pulse train waveform with the pump power, respectively. In the pumping regime where pulsation takes place, several chaotic regimes exist as if sandwiched by periodic regimes. Generally speaking, together with an increase in pump power, the pulse tends to exhibit ringing and its waveform becomes more and more complex. Chaotic regimes appear between different-type periodic pulsation regimes. For example, a chaotic solution exists in a transient region between the single pulse regimes, which appear immediately after the Casperson's instability (Fig. 1.9(f)), and the double pulse regime (Fig. 1.9(e)) as shown in Fig. 1.8. Also, the choatic solution of Fig. 1.9(d) appears in the process from the "doubled pulse" train to the "burst" train with ringing (Fig. 1.9(c)). As shown in Fig. 1.8, the dependence of the pulsation frequency on pump power agrees well with the experiment. Fig. 1.10 shows

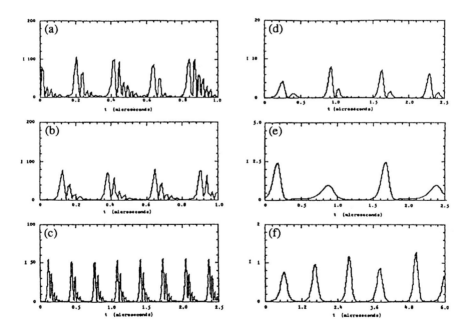

Figure 1.9: Theoretical pulsation waveform for a single-mode xenon laser duned to the line center tuning for different pump parameter values (a) $\alpha/\alpha_{th}^{(1)} = 3.0$ (chaos) (b) $\alpha/\alpha_{th}^{(k)} = 2.5$ (double burst) (c) $\alpha/\alpha_{th}^{(1)} = 2.0$ (chaos of single burst) (d) $\alpha/\alpha_{th}^{(1)} = 1.2$ (chaos of double pulse) (e) $\alpha/\alpha_{th}^{(1)} = 1.1$ (double pulse) and (f) $\alpha/\alpha_{th}^{(1)} = 1.06$ (chaos of single pulse). Horizontal unit is μs.

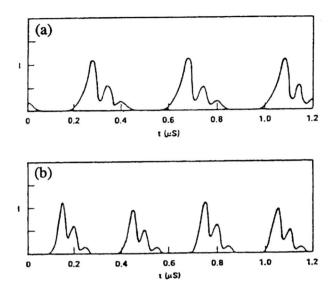

Figure 1.10: Experimental plots of output pulsations: (a) period-1 pulsation, (b) period-2 pulsation.

the output waveform observed with a Xe laser. If this waveform is compared with Fig. 1.9 (b), one can see the beautiful agreement between the theory and the experimental results. The fact that theory and experiment agree so well in such a complicated system as the gas laser is indeed amazing. However, unfortunately, complete physical understanding has yet to be achieved on the nonlinear dynamical mechanism which gives rise to such a variety of bifurcation processes.

1.3.2 Observation of three universal routes to chaos in a detuned system

The bifurcation route towards chaos was studied in detail by Gioggia and Abraham in a different FIR laser [10]. Figure 1.11 shows power spectra indicating typical bifurcation routes observed in a He-Xe laser. Here, three universal bifurcation routes were demonstrated for different lasing parameter regions.

[1] In the case (i), the main frequency f first appears. Next, subharmonic peaks at frequencies $f/2, f/2^2, f/2^3, \ldots$ appear, indicating successive period doubling routes.

[2] As for the case (ii), the main frequency f_1 appears at first. Next a second frequency f_2, which is incommensurate with f_1, appears ((b)). Note

1.3. INHOMOGENEOUSLY BROADENED SINGLE-MODE LASER

(i)
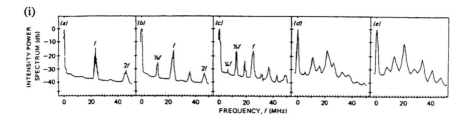

(ii)
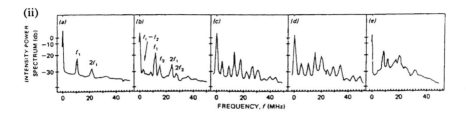

(iii)
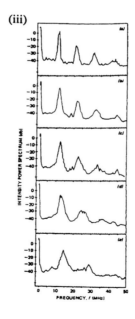

Figure 1.11: (i) Power spectra (resolution 1 MHz) of laser output intensity for different detunings showing period-doubling route to chaos. (ii) Power spectra (resolution 1 MHz) of laser output intensity showing quasi-periodic route to chaos for different values of the discharge current (laser gain variations) for 175-mTorr Xe and 0.3-Torr He. (iii) Intermittency featuring a broadening of pulsation spectral peaks (resolution 100 kHz) with increasing discharge current for 175-mTorr Xe and no helium.

that combination tones $n_1 f_1 + n_2 f_2$ are also formed. Together with an increase in pump power, the ratio $f_1 : f_2$ are locked in an integer ratio ((c)) and the spectrum broadens with a transition to chaos ((d)(e)). Consequently this phenomenon shows a typical T^2 locking $\to$ chaos route, e.g., quasiperiodicity.

[3] Additionally, in the case (iii), the bifurcation route resulting from intermittency is seen. With an increase in laser power, the spectrum gradually broadens, leading to chaos.

In all three cases, the bifurcation occurs when the cavity resonance frequency is detuned from the atomic line center ω_0. This fact strongly suggests that the effect of nonlinear refractive index stemming from the frequency detuning plays an essential role in bifurcation phenomena. Several numerical simulations have been carried out and three bifurcation routes have been reproduced by introducing the frequency detuning. However, correspondence between experimental and theoretical parameters is not always good. The clear physical interpretation of bifurcations is an important issue remaining for the future study.

1.4 Nonlinear Dynamics and Chaos in Semiconductor Lasers

Chaos in single-mode lasers discussed in the previous sections results from the coherent interaction between electric field and material field, in which the governing equations are described by electric field, population inversion and polarization. However, these types of laser chaos have been restricted to the rather nonpractical far-infrared lasers (so-called class C lasers). Thus, the question arises as to whether such chaotic behaviors take place in more practical laser devices, such as CO_2, solid-state, and semiconductor laser diodes (LDs). In such lasers, polarization decay time is much shorter than other time scales involving lasing. As a result, polarization dynamics are adiabatically eliminated in Maxwell-Bloch (MB) laser equations. This implies that MB instability can never been expected in ordinary situations because of the lack of degrees of freedom. This essential difficulty can be removed of course by introducing other degrees of freedoms in the form of external modulation, light injection, introduction of saturable absorber, external feedback and so on. Indeed, the period-doubling leading to chaos featuring relaxation oscillations was demonstrated in modulated CO_2 [11] and solid-state lasers [12]. [The first observation of laser chaos was reported in a modulated Nd:YAG laser in 1970 as mentioned in Preface].

Among these lasers, which are often categorized as class B lasers, LDs have an important inherent characteristic which cannot be expected in other lasers. That is an anomalous dispersion effect at the lasing frequency and the free-carrier plasma effect. These effects result in carrier-density dependent refractive index, which is expressed by the so-called α-parameter.

A resultant large $\chi^{(3)}$ nonlinearity gives rise to a variety of nonlinear phenomena in LDs in addition to the lasing action itself. This section reviews various nonlinear phenomena in LDs reported so far, paying special attention to experimental demonstrations and their physical interpretations.

1.4.1 Inherent $\chi^{(3)}$ nonlinearity in laser diodes (LDs)

The gain spectrum of LDs is asymmetric with respect to the gain peak frequency reflecting the band structure. Furthermore, the loss for lasing field guided in the active layer is also frequency dependent. With these two effects combined, the oscillation frequency is detuned from the gain spectrum peak towards lower frequencies. This results in a peculiar anomalous dispersion effect at the lasing frequency based on the Kramers-Kronig relationship as is described by Eq. (1.9). This effect, together with the many-body free-carrier plasma effect, results in the refractive index (n) dependence on carrier density (N) and can be characterized by the following quantity

$$\alpha = -(4\pi/\lambda)(\partial n/\partial N)/(\partial G/\partial N). \qquad (1.41)$$

For ordinary LDs, α ranges from 3 to 7, whereas $\alpha \approx 0$ for other laser systems.

Using this quantity (e.g., so-called α parameter), the complex susceptibility of LDs can be expressed as

$$\chi(N) = (n/k_0)(-\alpha + j)G(N), \qquad (1.42)$$

where k_0 is the wavenumber in vaccum. If the gain (G) is expanded into Taylor's series with respect to the laser intensity I, the third-order susceptibility $\chi^{(3)}$ is given by

$$\chi^{(3)} = (n^2 \epsilon_0 c/k_0)(\alpha - j)(G_0/I_s), \qquad (1.43)$$

where G_0 is the unsaturated gain, assuming $I \ll I_0$ (I_0: saturation intensity). Assume InGaAsP lasers, for example, $\chi^{(3)}$ is on the order of 6×10^{-6} esu. This is so large that only the order of 100 kW/cm^2 optical intensity is required to achieve π shift in the 300 μm-length LD.

Such a nonlinearity results in a strong (parametric) four-wave mixing process which gives rise to the peculiar mode interaction in LDs. If we introduce another degree of freedom into LDs, such as external modulation and/or an external mirror, qualitatively different instabilities from Maxwell-Bloch instabilities, featuring longitudinal mode spacing frequency and relaxation oscillation frequency, take place resulting from the third-order nonlinear interaction among oscillating modes.

1.4.2 Modulated extended cavity laser diodes

The active mode-locking techique of lasers is widely used to control multimode oscillations to produced optical pulse trains (phase-locking) or frequency-modulated output (FM-laser operation). Such operations were well-studied in the 60's, but theoretical investigations so far are applicable only to class A lasers (e.g., gas and dye lasers) in which polarization and population dynamics are adiabatically eliminated. In directly modulated LDs, on the other hand, population (carrier) dynamics play an essential role in mode-locking behavior. In particular, nonlinear mode interaction resulting from the third-order process (population pulsation) results in unique mode-locking dynamics.

A. Subharmonic resonance FM-laser operation

Mukai *et al.* observed peculiar FM-laser operations in a directly-modulated extended cavity LD [13]. The experiments were performed using a 1.52-μm InGaAsP buried heterostructure LD, whose facet facing the external mirror was antireflection-coated (0.1% reflectivity), and a flat mirror. The cavity length was $L = 28$ cm. The extended cavity LD used in this particular experiment exhibited a TM-polarized pure single-frequency operation at a wide injection current range of up to $15 \times I_{th}$ (threshold).

When the laser was modulated by a small sinusoidal injection current at frequency f_m close to the longitudinal mode spacing frequency $f_c = c/2L$, an FM output was achieved as shown in Fig. 1.12(a), while a weak spurious amplitude modulation (1–10%) was superimposed due to direct current modulations. This observation is easily explained. When the injection current is modulated, carrier density is modulated accordingly. This results in the modulation of gain as well as phase at f_m resulting from the carrier-density dependent refractive index. Consequently, the LD tends to be phase modulated and FM sidebands are created around the free-running mode f_0. These Bessel function components are then resonantly enhanced because of the presence of cavity resonance as shown in Fig. 1.13, although weak AM components are superimposed. Such an FM-laser operation is not non-trivial and can be understood well based on Harris-McDuff's traditional FM-laser theory [14].

In the case of LDs, however, FM-laser oscillations are also resonantly achieved for subharmonic modulation as shown in Figs. 1.12(b) and 1.13. Such subharmonic resonance phenomena inherent in LDs, which can never be observed in ordinary $\alpha = 0$ lasers, are explained as follows. (See Fig. 1.14.) Assuming that the LD is modulated at $f_c/m (m = 2)$, nonlinear interaction between the free-running mode f_0 and its sidebands ($f_0 \pm f_c/m$) takes place. These weak sideband waves act as probe beams in the nearly degenerate four-wave mixing scheme. In short, population pulsation occurs

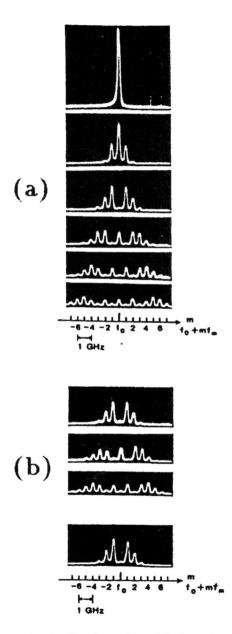

Figure 1.12: Output spectra of a directly modulated single mode extended cavity LD for different modulation currents. (a) Fundamental FM-laser operation, (b) Subharmonic resonance FM-laser. Upper three traces are for $f_m = (1/2)f_c$ and the bottom is for $f_m = (1/3)f_c$.

26   CHAPTER 1. PROLOGUE TO NONLINEAR DYNAMICS

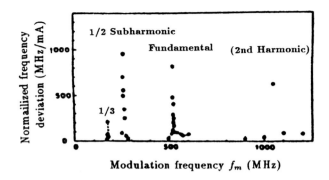

Figure 1.13: Frequency response of FM-laser operations. Frequency deviation per unit sinusoidal current amplitude as a function of modulation frequency.

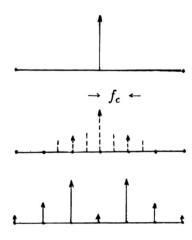

Figure 1.14: Physical interpretation of subharmonic resonance FM-laser operations in LD.

at the frequency f_c/m and combination-tone polarizations (signals) are created as a result of third-order nonlinearity. At this moment, however, the only electric-field components located at cavity resonance frequencies are considered to be resonantly enhanced. Consequently, the population pulsation component at f_c becomes dominant and causes phase modulation in the LD at f_c, giving rise to fundamental FM-laser operations whose spectra coincide with Fig. 1.12(a).

1.4. NONLINEAR DYNAMICS AND CHAOS IN SEMICONDUCTOR LASERS

B. Devil's staircase mode-locking

A peculiar instability featuring similar subharmonic resonances also appears in the case of phase-locking experiments in an extended cavity LD. Baums et al. prepared an extended cavity LD which operates in multiple longitudinal modes separating f_c and is modulated by an injection current [15]. Figures 1.15(a) show typical examples of pulse trains for fundamental phase-locking $f_m = f_c$ and $f_m = (2/5)f_c$. Corresponding rf power spectra are shown in Fig. 1.15(b). In this experiment, pulse patterns belonging to f_m/f_c equal to the fractions $n + p/q$ with $n = 0, 1$ and $p = 1$ to q for q up to 6 were observed. For example, Fig. 1.15(b) corresponds to the Farely fraction with $n = 0$, $p = 2$, and $q = 5$. Plotting the Farely fractions of the frequency locked states as the winding numbers against modulation frequency yields the upper curve in Fig. 1.16, revealing a devil's staircase. The gaps between the stairs refer to quasiperiodicity.

A rigorous theoretical explanation has yet to be obtained, but the nonlinear wave-mixing process resulting from population pulsation among lasing fields and excited sideband fields might play an essential role in p/q phase-locking similarly to the FM-laser operation.

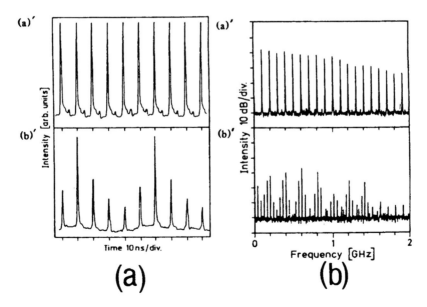

Figure 1.15: (a) Output pulse trains from a directly modulated multimode extended cavity LD for (a)' $f_m = f_c$ (fundamental mode locked) and (b)' $f_m = (2/5)f_c$ (frequency locked). (b) Corresponding rf spectra for (a)' and (b)' in (a).

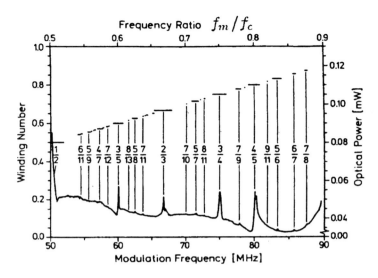

Figure 1.16: Measured winding number (upper curve belonging to left vertical axis) and emitted optical power (lower curve belonging to right vertical axis) as a function of the modulation frequency, showing devil's staircase locking.

1.4.3 Autonomous instabilities in composite cavity laser diodes

Dynamic instabilities in a laser diode coupled to an external cavity resulting from the third-order nonlinearity expressed by the α-parameter was initiated by Lang and Kobayashi in 1980 [16]. Numerous publications concerning feedback-induced instabilities in LDs have been reported so far and they are classified into several classes depending on the operating conditions. In this subsection, typical examples are summarized.

A. Self-sustained relaxation oscillation in a weakly coupled system

Influences on the LD oscillation properties of weak external optical feedback from an external reflector placed few centimeters from the output LD facet were first investigated by Lang and Kobayashi, who observed self-sustained relaxation oscillations [16] in the regime that an external cavity length is as short as 1–2 cm. The relaxation oscillations at the frequency $f_r = (1/2\pi)\sqrt{(w-1)/\tau\tau_p}$ (w: pump power/threshold, τ: population lifetime, $\tau_p = 1/2\kappa$: photon lifetime) are inherent in class-B lasers, which result from the interplay between photons and population inversion in rate equations, with the dynamics of polarization being eliminated adiabatically from Maxwell-Bloch laser equations. [This point will be described in Chapters **2-4**].

Figure 1.17 shows the transient output response to a current pulse at

1.4. NONLINEAR DYNAMICS AND CHAOS IN SEMICONDUCTOR LASERS

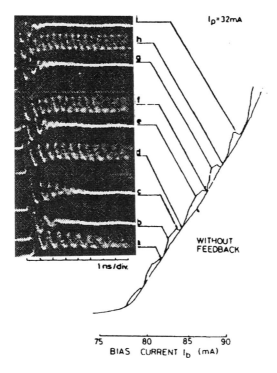

Figure 1.17: The transient output response to current pulse at various points on the undulated $L-I$ curve of a weakly-coupled external cavity LD.

various points on the undulated $L-I$ curve. Self-sustained relaxation oscillations appear near the bottoms of undulated output. The experimenters introduced a delayed feedback term into the lasing field dynamics. Combined with the carrier dynamics, they derived the following generalized Van der Pol equations involving the lasing field $\tilde{E}(t)$ and the population inversion $N(t)$ with the α-nonlinearity incorporated.

$$dN(t)/dt = P - N(t)/\tau - G(N)|E(t)|^2, \qquad (1.44a)$$

$$dE(t)e^{i\omega_s t}/dt = [i\omega_{LD}(N) + (1/2)(G(N) - 1/\tau_p)]E(t)e^{i\omega_s t}$$
$$+\zeta E(t - t_D)e^{i\omega_s(t-t_D)}, \qquad (1.44b)$$

where $G(N) = 1/\tau_p + (\partial G/\partial N)(N - N_{th})$. Here, ω_{LD} is the diode cavity resonant frequency, ω_s is the compound-cavity eigenmode frequency, $P = J/ed$ (J: injection current density, e: electric charge, d: active layer thickness), N is the carrier density, N_{th} is the threshold carrier density, $G(N)$ is the gain, τ is the carrier lifetime, τ_p is the photon lifetime, $\zeta = (1 - r_0^2)(cr_1/2n\ell_1 r_0)$ is the feedback parameter (r_0: output facet amplitude reflectivity, r_1: feedback mirror amplitude reflectivity, ℓ_1 is the LD length) and t_D is the delay time.

They carried out the linear stability analysis for the stationary solutions of Eq. (1.44) and showed that dynamical instability featuring sustained relaxation oscillations occurs at valleys of undulation curve as shown in Fig. 1.17 in a weakly coupled regime, e.g., $\zeta \simeq 2 \times 10^8/s$.

The physical interpretation is as follows. In such a weakly-coupled system (Fig. 1.18(a)), the relationship between the diode cavity mode

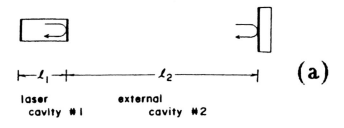

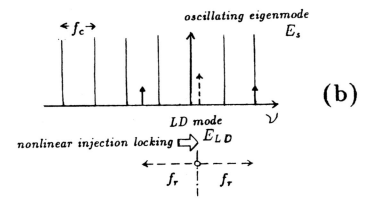

Figure 1.18: (a) Conceptual model of an LD coupled to an external reflector. (b) Frequency arrangements of diode cavity mode E_{LD} and external cavity oscillating modes E_s, and nonlinear injection locking process among these modes.

1.4. NONLINEAR DYNAMICS AND CHAOS IN SEMICONDUCTOR LASERS

E_{LD} and oscillating external cavity eigenmodes E_s is considered to be like Fig. 1.18(b). Let us consider the interaction between two nearby modes. If the frequency difference between two fields is within the lock-in range, the interference condition is satisfied and E_s is *injection-locked* to E_{LD}. At this point, output increases exhibiting the undulation peaks in Fig. 1.17. Suppose $\Delta\Omega = \omega_s - \omega_{LD}$ to be negatively too large for the intereference condition to be optimum. The intensity increase in the transient relaxation oscillation waveform reduces the carrier density due to the saturation effect, with decrease in the gain and increase in the refractive index in the diode cavity according to Eq. (1.9) or Eq. (1.41). The refractive index increase shifts ω_{LD} toward ω_s and thus improves the interference condition, resulting in the further increase in the intensity of E_s. Such a positive feedback effect balances with the damping, and self-sustained relaxation oscillation takes place, when $\Delta\Omega$ is sufficiently negative, while it is strongly damped out when $\Delta\Omega$ is positive. (See Fig. 1.18(b).) Such an asymmetric interaction will be described again in C in this subsection.

B. Quasiperiodic route to chaos in a weakly coupled system

Mørk et al. observed competition between the above-mentioned relaxation oscillations and the instability corresponding to the inverse transit time in an external cavity, i.e., f_c [17] in the regimes that an external cavity length is increased. Measured intensity spectrum (a) and frequency ratio f_r/f_c versus feedback level (b) are shown in Fig. 1.19. The example (a) demonstrates a frequency-locked state where all peak frequencies are multiples of f_c, and $f_r/f_c = 5$.

In Fig. 1.19(b), a strong relaxation oscillation peak at R indicating sustained relaxation oscillation appears in regime I. In regime II, the amplitude of C (i.e., f_c) increases strongly, which indicates quasiperiodicity. Regime III is characterized by a plateau in f_r/f_c at the integer 5

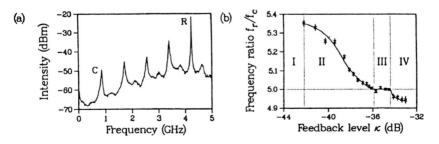

Figure 1.19: Subharmonic resonance and quasiperiodicity in a weakly-coupled composite cavity LD. (a) Power spectrum indicating $f_r/f_c = 5$ locking. (b) Frequency ratio f_r/f_c as a function of feedback level.

(frequency-locking). The range of frequency locking depends strongly on the laser parameters, and in some cases the quasiperiodic attractor evolves into a chaotic state in regime IV. Such a subharmonic resonance and the quasiperiodic route to chaos might be interpreted in terms of the breakup of a 2-torus similarly to the case of active phase-locking discussed in **1.4.2 B**.

C. Subharmonic resonance cascade below the kink in a strongly coupled system

As a result of the anomalous dispersion effect described in **1.4.1**, peculiar mode coupling, which cannot be seen in other lasers, takes place in LDs. If a weak probe field E_a at angular frequency ω_a is introduced into a LD cavity, which is saturated by a strong laser radiation E_s (ω_s), the carrier density pulsation takes place at a beat frequency between the two fields. The carrier density pulsation creates a moving grating within LD and results in the amplitude as well as phase modulation through α-parameter. As a consequence, dynamic scattering of the strong field toward the weak probe field, whose amount is proportional to $E_a|E_s|^2$, occurs.

The parametric gain for the probe beam E_a due to such process is given by [18]

$$\delta g \propto (\alpha\Delta\Omega + 1)/(\Delta\Omega^2 + 1), \qquad (1.45)$$

where $\Delta\Omega = (\omega_a - \omega_s)/\Gamma$ (Γ: stimulated recombination probability). Figure 1.20 shows the additional gain δg for E_a as a function of frequency separation. It is concluded from this figure that E_a obtains the parametric gain when it locates at the lower frequency side with respect to the strong lasing field. Physically, the scattered field from E_s is added constructively to E_a located in the lower frequency side, while it destructively interferes with E_a when E_a locates in the higher frequency side [18]. Such asymmetrical coupling in LDs has been extensively studied and been confirmed experimentally, in nearly degenerate four-wave mixing scheme [19].

Mukai and Otsuka observed subharmonic resonance and chaotic relaxation oscillations in an LD strongly coupled to an external mirror resulting from the parametric four-wave mixing, by changing the mirror tilt and setting f_c to around the relaxation oscillation frequency f_r [20]. [This was the first clear demonstration of chaos in a compound-cavity laser diode.] Figure 1.21 shows (a) $L-I$ curve, (b) oscillation spectra, (c) output waveforms, and (d) corresponding power spectra. This experiment confirmed that the relaxation oscillation noise peak at f_r is moved around f_c by changing the injection current slightly.

In strongly coupled LDs, in general, the lasing threshold decreases to a great extent like (a) and below the 'kink', the system acts as a composite cavity laser oscillator. The instabilities shown in Fig. 1.21 are observed below the 'kink'. In contrast to weakly-coupled LDs, frequencies of composite cavity eigenmodes ω_s with optimum gain, which are derived from

1.4. NONLINEAR DYNAMICS AND CHAOS IN SEMICONDUCTOR LASERS 33

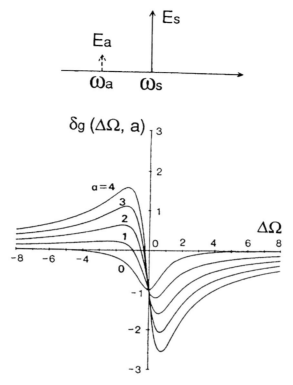

Figure 1.20: Aymmetrical interaction between the lasing field E_s and the probe field E_a in LDs resulting from the $\chi^{(3)}$ nonlinearity, e.g., α parameter.

Lang-Kobayashi Equation (1.44) with a field continuum condition at the output facet, largely shift from the laser diode mode frequency ω_{LD} due to the reduction of threshold carrier density. Therefore, only composite cavity eigenmodes involve dynamics below the kink region. Furthermore, if a small *tilt* of the external mirror is introduced, eigenmodes are shown to bifurcate into two mode groups, as indicated by the solid and open circles which have different gains like *double combs* shown in Fig. 1.22(a) [21]. The frequency separation between two mode groups, i.e., $\omega_s - \omega_a$, changes with the tilt. The mode group which has lower gain cannot break into oscillation, but the modes are considered to act as amplified spontaneous emission (ASE) modes reflecting the large gain of LDs.

The essential features are thought to be extracted by three mode interactions, namely, two adjacent lasing modes E_{s1} (frequency f_{s1}), $E_{s2}(f_{s2})$ and the ASE mode $E_a(f_a)$. Due to the parametric four-wave mixing be-

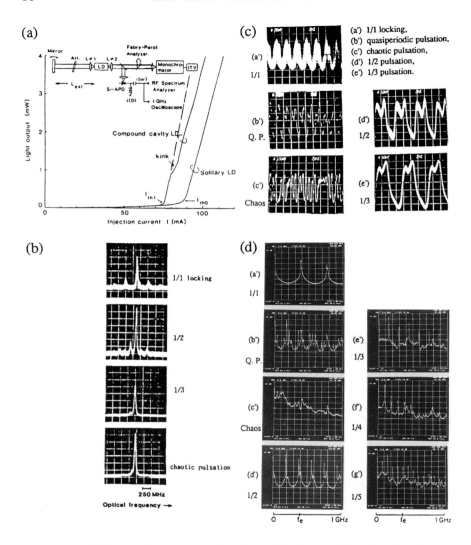

Figure 1.21: Subharmonic resonance relaxation oscillations and chaotic relaxation oscillations in a LD strongly coupled to a tilted external mirror below the kink regime. (a) $L - I$ curve, (b) Oscillation mode spectra, (c) Output waveforms, (d) Corresponding power spectra (10dB/div).

tween E_a and E_{s1}, a sideband field $E_d(f_d)$ is created near the E_{s2} field. This scattered field then acts as an injection light to E_{s2} as shown in Fig. 1.22(b). When the injection locking is established, these three modes involved are resonantly excited as in Fig. 1.22(c). Here, the relaxation oscillation plays

1.4. NONLINEAR DYNAMICS AND CHAOS IN SEMICONDUCTOR LASERS 35

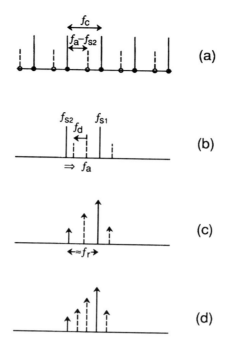

Figure 1.22: Frequency arrangement of eigenmodes of tilted external mirror cavity LDs and physical interpretation of subharmonic resonance relaxation oscillations.

an important role. In other terms, if the relaxation oscillation frequency is close to f_c (longitudinal mode spacing freqency), the beat note at $f_{s1} - f_d$ can excite sustained relaxation oscillation which appears with 1/2 subharmonic resonance. When the mirror tilt is changed, $f_{s1} - f_a$ changes [21] and the fractional resonance $f_{s1} - f_a = f_r/n$ (n: integer) is expected. ($n = 3$ for Fig. 1.22(d)). As is justified by the experiment, quasi-periodic and chaotic relaxation oscillations appear when the fractional locking condition is not satisfied.

Otsuka and Mukai [22] showed that experimentally observed waveforms and power spectra are reproduced by the numerical simulations of coupled generalized Van der Pol equations for three electric fields and population inversions [See *Supplement 1*]. The result is shown in Fig. 1.23, where $(f_a - f_{s2})/f_c$ is changed. The experimental results well correspond to simulated results, and the instability scenario in terms of parametric four-wave mixing and resonant excitation of subharmonic sustained relaxation oscillations are confirmed to explain the experimental results almost completely.

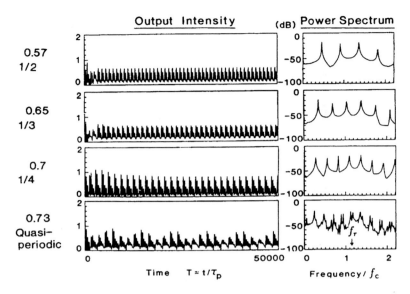

Figure 1.23: Simulated waveforms and corresponding power spectra for various frequency differences between E_a and E_{s2}. Note that the relaxation oscillation frequency f_r is located near f_c.

D. Coherence collapse above the kink in a strongly coupled laser diode

Above the kink in strongly coupled composite cavity LDs with a distant reflector, a qualitatively different instability takes place. This instability is referred to as "coherence collapse" [23]. Figure 1.24 shows (a) $L - I$ and power spectra, (b) output waveforms and (c) the Poincaré plot reported by Sacher et al. [24]. Above the kink, a broader noise peak appears around f_c with lower frequency noise components extended. From the characteristic distribution of the mean time of regular behavior $\langle T \rangle \simeq \epsilon^{-1}$ ($\epsilon \equiv (I - I_{th,s})/I_{th,s}$; $I_{th,s}$: threshold current of the solitary LD) and Poincaré plot, they claimed the occurence of inverse type-II intermittency.

It should be careful, however, that output waveforms featuring low-frequency intensity drops are obtained by low-pass filtering the signal with a cutoff frequency of 1 GHz, hiding any faster dynamics resulting from relaxation oscillations. Most recently, Fisher et al. reported the experimental demonstration of irregular fast pulsing within the coherence collapse by the streak camera measurements [25a)]. Irregular picosecond light pulses are born from a Hopf instabilities (e.g., Lang-Kobayashi's sustained relaxation oscillations described in A) and they explained this faster dynamics in terms of *chaotic itinerancy* among attractor ruins of coexisting eigenmode states (constructive interference solutions of (1.44))[25a)]. Numerical sim-

1.5. PERIOD-DOUBLING BIFURCATION

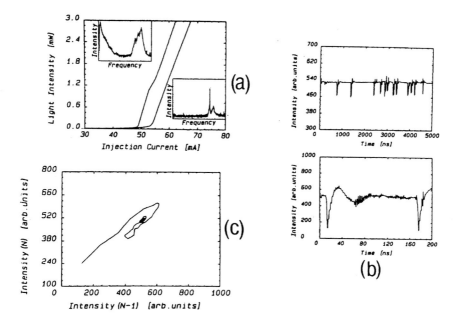

Figure 1.24: Inverse type-II intermittency above the kink regime. (a) $L - I$ curve and power spectra, (b) Output waveforms, (c) Poincaré plot.

ulations indicated that intensity drops are associated with the self-induced switching among eigenmode attractor ruins *via antimodes*, which referred to as the destructive interference solutions of Lang-Kobayashi equations (1.44) exhibiting a saddle-node instability.

The origin of coherence collapse may arise from the mixed effect of spontaneous emission noise (stochasticity) and dynamical instability described here. This is still a challenging problem [25b)].

1.5 Period-Doubling Bifurcation in Nonlinear Passive Resonator

1.5.1 Delay-differential equation model

In this chapter so far, prototypical examples of chaotic laser dynamics in class C single-mode lasers (e.g., far-infrared lasers) described by Maxwell-Bloch laser equations and in class-B lasers (e.g., semiconductor laser diodes) described by generalized Van der Pol laser rate equations, in which polarization dynamics are adiabatically eliminated, are reviewed. In this section, instabilities and chaos in nonlinear "passive" resonators with injected signal are reviewed, in which the laser medium is replaced by nonlinear refractive index media.

Ikeda introduced a nonlinear passive resonator modelled by delay-differential equations [D]. The original resonator scheme is illustrated in Fig. 1.25, in which a nonlinear medium is located within a ring resonator and the external light is injected into the resonator.

Suppose that the polarization field follows the electric field change adiabatically, the Maxwell-Bloch equations (1.6)–(1.8), in which the left-hand side Eq. (1.6) is re-written as $c\partial \tilde{E}/\partial z + \partial \tilde{E}/\partial t$ to include a propagation effect, are integrated with respect to the spatial variable z under the boundary condition at the mirror 1 and 2

$$\tilde{E}(t,0) = \sqrt{1-R}E_I(t) + R\tilde{E}(t-(L-\ell)/c, \ell), \qquad (1.46)$$

where $\tilde{E}_I$ is the amplitude of incident laser light, L is the cavity length and ℓ is the medium length and R is the reflectivity of the mirror. Then, the Maxwell-Bloch equations are tranformed into the following delay-differential equations.

$$E(t) = A + BE(t-t_R)\exp(\phi/\Delta)\exp i[\phi(t)-\phi_0], \qquad (1.47)$$

$$\gamma^{-1}d\phi(t)/dt = -\phi(t)+\text{sign}(\Delta)|E(t-t_R)|^2(1-e^{2\phi/\Delta-\alpha\ell})/(1-e^{-\alpha\ell}). \qquad (1.48)$$

Here, $E(t) \equiv [k|n_2|(1-e^{-\alpha\ell})/\alpha]^{1/2}\hat{E}(t,0)$ is the normalized electric field at the entrance of the medium, where α is the absorption coefficient, k is the wavenumber, $n_2 = \chi^{(3)}/2\epsilon_0 n_0$ is the nonlinear refractive index defined by $n = n_0 + n_2|\tilde{E}|^2$ (n_0 is the linear refractive index of the medium). Variable ϕ denotes the phase shift suffered by the electric field in the medium and ϕ_0 is the linear phase shift. t_R is the round-trip time along the optical path and $\Delta \equiv (\omega_0-\omega)/\gamma$. For a brevity, the notation change $\gamma_\parallel \to \gamma$ is carried

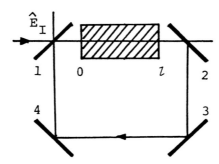

Figure 1.25: A conceptual model of the nonlinear resonator irradiated by a coherent laser light.

1.5. PERIOD-DOUBLING BIFURCATION

out. The parameter $A \equiv [(1-R)k|n_2|(1-e^{-\alpha\ell})/\alpha]^{1/2}|\tilde{E}_I|$ is the normalized incident field amplitude and the parameter $B \equiv Re^{-\alpha\ell/2}$ characterizes the dissipation of the electric field in the cavity. A detailed derivation of Eqs. (1.47)–(1.48) is given in the textbook [E].

In the *dispersive limit*, i.e., $\Delta \gg 1$, Eqs. (1.47)–(1.48) are simplified to

$$E(t) = A + BE(t - t_R)\exp[i(\phi(t) - \phi_0)], \tag{1.49}$$

$$\gamma^{-1}d\phi(t)/dt = -\phi(t) + |E(t - t_R)|^2, \tag{1.50}$$

These model equations (1.49)–(1.50) are simplified further in the limit of large dissipation, i.e., $B \ll 1$. If A^2B is kept constant, we can eliminate the electric field and the following simple model equation is obtained.

$$\gamma^{-1}d\phi(t)/dt = -\phi(t) + A^2 F(\phi(t - t_R)), \tag{1.51}$$

where $F(\phi) = 1 + 2B\cos(\phi - \phi_0)$

1.5.2 Ikeda instability

The delayed feedback common in the models described in the previous subsection causes an instability leading to self-oscillation with the period of two times the delay time, i.e., $2t_R$. Its essential feature is descibed by the difference equation, e.g., Ikeda map:

$$\phi(t) = A^2 F(\phi(t - t_R)), \tag{1.52}$$

which is obtained assuming $\gamma t_R \gg 1$ in the model equation (1.51).

Let us consider the bifurcation structure leading to chaos based on this simplest model. Fig. 1.26 shows how the graphical solution changes with an increase in the input intensity $\mu \equiv A^2$, where (a) indicates the period $2t_R$ solution, while (b) indicates the period-doubled $4t_R$ solution. Such period-doubling $2t_R \to 4t_R \to 8t_R \cdots$ occurs successively as μ exceeds the bifurcation thresholds $\mu_1, \mu_2, \mu_3, \ldots$. These bifurcation points converge exponentially to a finite value μ_F. When μ exceeds μ_F, the chaotic state appears. This is nothing more than the Feigenbaum's period-doubling bifurcation route to chaos. A typical example of bifurcation diagram is shown in Fig. 1.27.

1.5.3 Physical interpretation of Ikeda instability

The physical interpretation of delay-induced instability, featuring the successive period-doubling bifurcation, is given in terms of four-wave mixing process [26].

The frequencies of cavity modes in resonance with a nonlinear optical resonator depend on the light intensity; they move in proportion to the

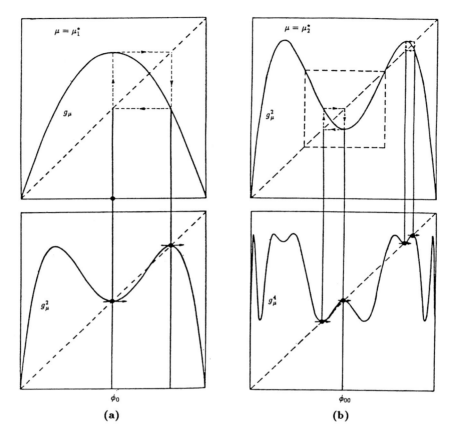

Figure 1.26: Graphical solution of the Ikeda map. (a) period-1 solution, (b) period-2 solution.

light intensity as $\omega_n = 2\pi n/t_R + \text{const.}|E|^2$ due to the presence of nonlinear refractive index n_2.

On the other hand, the third-order nonlinearity, which is the origin of nonlinear refractive index, enables two photons of frequencies ω_1 and ω_2 to create new photons of frequencies ω_3 and ω_4 through the scattering process, where the energy conservation condition $\omega_1/c + \omega_2/c = \omega_3/c + \omega_4/c$ is fulfilled. This process is responsible for a parametric four-wave mixing gain.

Now, let us consider the case that the nonlinear optical resonator is irradiated by a coherent laser light with frequency ω. As a result of four-wave mixing, two photons are created symmetrically with respect to ω. As long as the frequencies of the two scattered photons do not coincide with the

1.5. PERIOD-DOUBLING BIFURCATION

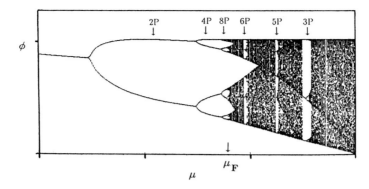

Figure 1.27: A typical example of period-doubling bifurcation in the Ikeda-map system.

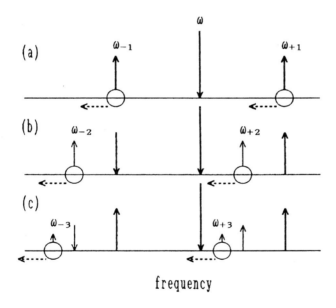

Figure 1.28: Physical interpretation of period-doubling bifurcation by four-wave mixing process.

frequencies of the cavity modes, the scattering process does not work effectively. When the light intensity is increased, cavity resonance frequencies shift as indicated by arrows. At the moment when cavity resonance frequencies $\omega_{\pm 1}$ locate symmetrically with respect to ω as shown in Fig. 1.28(a), the scattered photons with frequencies $\omega_{\pm 1}$ are amplified. As a result, the oscillation with the period $2t_R$ takes place (Fig. 1.28(b)). When the light intensity is increased further up to the point shown in Fig. 1.28(b), scattered photons with frequencies $\omega_{\pm 2}$, which satisfy the condition of $\omega_{-1} + \omega = \omega_{-2} + \omega_{+2}$, are amplified and the oscillation with the period $4t_R$ occurs at this moment (Fig. 1.28(c)). This process repeats and the successive period-doubling bifurcation takes place.

1.5.4 Experimental observation of Ikeda instability

The first experimental observation of Ikeda instability was done by utilizing a hybrid electro-optic bistable device, in which feedback is introduced by using an electrical circuit with a delay line [27]. Period-doubling bifurcation and chaos were experimentally observed also in the all-optical system, which is compatible with the model equation (1.52). The idea is to construct a nonlinear ring resonator using a single-mode optical fiber as a nonlinear refractive index medium. A mode-locked pulse train with a period t_R is introduced into the fiber ring resonator [28]. By increasing the pulse intensity, period-doubling and chaotic modulation of pulse train were demonstrated. Results are shown in Fig. 1.29. The Ikeda oscillation and its subharmonics were also observed by Harrison *et al.* using an all-optical resonator containing a NH_3 subjected to "smooth" input pulses from a CO_2 TEA laser [29].

Bistability and Ikeda type instability in resonant-type bistable laser diode amplifiers are proposed by Otsuka and Iwamura [30] and extremely-low power bistable amplification is experimentally demonstrated on cw basis [31]. This is due to the α nonlinearity in LD's and this system is a promising candidate for demonstrating Ikeda instability. Bistability and chaos in LD's with injected signals are also demonstrated [32–33].

1.5.5 Bifurcation and dynamical memory in the Ikeda-map systems

In this subsection, applicability of Ikeda map to large-capacity *dynamical memory* is described briefly.

A. Tree-structure and isomers

If an optical beam, which is injected into the dispersive nonlinear resonator, reaches a threshold μ_1, the periodic oscillation with a period $2t_R$ (t_R is the roundtrip time of the cavity) appears as shown in **1.5.2**. As the input intensity increases and exceeds thresholds $\mu_2(<)\mu_3 \cdots$, the period-doubling bifurcation leading to chaos takes place. Just above the first bifurcation, i.e., μ_1, odd-order higher harmonics whose period is $2t_R/3, 2t_R/5, \ldots$ can also be realized [34]. At higher bifurcation points $(\mu_2, \mu_3, \ldots)$, multiple

1.5. PERIOD-DOUBLING BIFURCATION

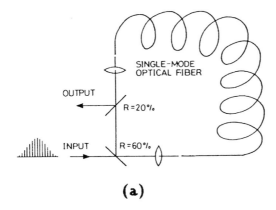

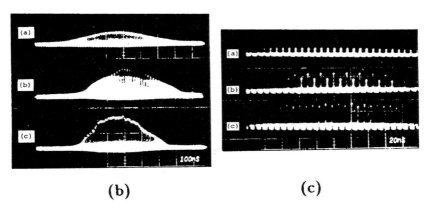

Figure 1.29: Experimental demonstration of Ikeda instability. (a) Experimental configuration, (b), (c) Period-doubling to chaos.

number of isomers are created from its mother structure. Assume N-th harmonics, then the fundamental section $I_n \equiv \{nt_R < t < (n+1)t_R\}$ can be devided into N subsections $I_l^{(1)}, I_l^{(2)}, \ldots, I_l^{(N)}$. If $\phi_l(\tau)$ in $\tau \in I_l^{(k)}$ ($k = 1, 2, \ldots, N$) is chosen to be one of the 2^m solutions $(\phi_1, \phi_2, \ldots, \phi_{2^m})$, these solutions are found to be stable solutions which satisfy the dynamics $\phi_{n+1}(\tau) = F(\phi_n(\tau))$. Figure 1.30 shows the isomers created from the third harmonic oscillation.

The bifurcation tree from various harmonics is depicted in Fig. 1.31. In general, the number of isomers is given asymptotically as $2^{N(M-1)}$, where M is the generation number. Therefore, if these isomers are assignable, the

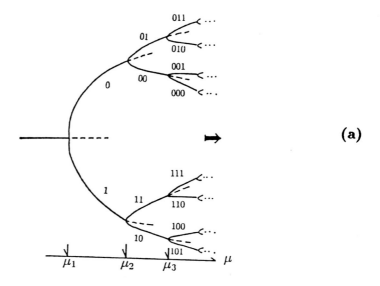
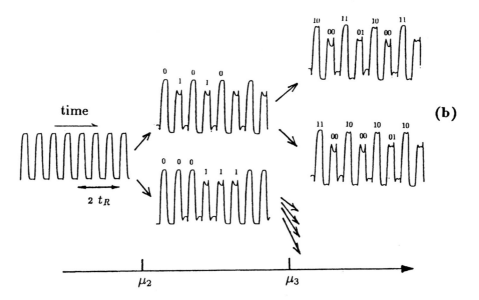

Figure 1.30: (a) Bifurcation procedure in an Ikeda resonator. (b) Isomers created from a third harmonic solution.

1.5. PERIOD-DOUBLING BIFURCATION

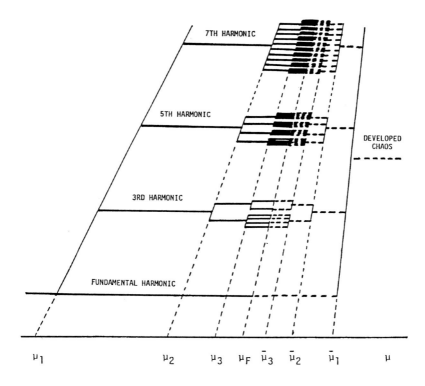

Figure 1.31: Tree structures of isomers.

memory capacity of this system is evaluated as $C = N(M-1)$ bits. This means that the bistable element, which possesses one bit memory capacity, obtains an extremely large memory capacity by introducing the delay. If N and M become large, the treatment based on the difference equation becomes invalid since the adiabatic elimination of the phase cannot be employed. In this case, the minimum memory capacity is evaluated to be $C = t_R \gamma$ bit from the original delay-differential equations. Assume $t_R = 1 ns$ and $\gamma^{-1} = 10 ps$, then $C = 100 bit$.

B. Assignment to desired isomers

In order to realize dynamic memory, arbitray isomer patterns should be assigned by external signal. If the bifurcation tree is known, the assignment to the desired isomers are easily established by adding the trigger signal, whose pattern is identical to the desired pattern, within an time interval of t_R. This modulation signal acts as a seed and the desired patterns are repetitively produced. Figure 1.33 depicts an example of the

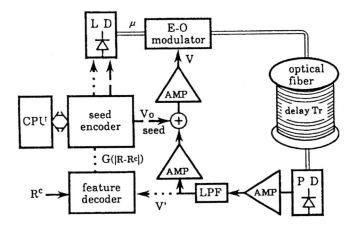

Figure 1.32: Experimental setup for dynamic memory. LD: laser diode, E-O: electrooptic modulator, PD: pin photo-diode, AMP: video amplifier.

assignment process to one of the third-generation ($M = 3$) isomers. By adding the trigger signal (seed) to the medium just before the bifurcation points, the desired pattern can be realized. Such a *seeded bifurcation switch* (SBS) was demonstrated in a hybrid optical system shown in Fig. 1.32, in which an optical fiber is introduced to control a delay time [35–36]. [For a fundamental delay-differential equation describing such an optoelectronic hybrid device, see *Supplement 2*]. In order to realize the switching between isomers, one must reset the system to the the first generation and then access to the new isomer. Therefore, the long access time is required in this system.

The basin of attraction for individual isomer becomes narrower exponentially with the generation number M and the transition between isomers occurs. In this regime, an erratic wandering between isomers takes place and chaos develops. The chaotic time series are expected to possess chacteristic features of isomers from which chaos was born. Therefore, the possibility arises in which the desired isomer can be assigned from chaotic sea by extracting the characteristic features and by feeding the control signal back to the system, until the error between the desired pattern and appearing pattern diminishes. Such an assignment process, namely *chaotic search* (CS), has also been demonstrated to be possible in the present system [35–36], although the long access time is required as well. The CS has also been experimentally demostrated utilizing a hybrid optical system [Fig. 1.32] as shown in Fig. 1.34, where evolutions of output and error signal ($|R - R_c|$) are shown.

1.5. PERIOD-DOUBLING BIFURCATION

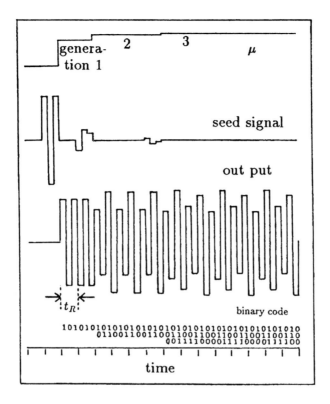

Figure 1.33: Assignment to a third generation isomer.

Supplement 1: Coupled Generalized van der Pol Laser Equations

Numerical simulations which reproduced subharmonic-resonance cascade leading to chaos in an laser diode strongly coupled to a tilted external mirror have been calculated by the following equations [22]:

$$dN(t)/dt = P - N/\tau - G(N)S, \quad (S.1)$$

$$dS_a/dt = \eta[G(N) - 1/\tau_p]S_a + \epsilon N/\tau, \quad (S.2)$$

$$dE_{sm}/dt = [G(N) - 1/\tau_p](E_{sm/2}) + qE_a \cos \Psi_{sm}, \quad (S.3)$$

$$d\Psi_{sm}/dt = \omega_{sm}(N) - \omega_a - q(E_a/E_{sm}) \sin \Psi_{sm}. \quad (S.4)$$

Here, $S_a \equiv E_a^2$, $S \equiv S_a + \sum_m E_{sm}^2$, $m = 1, 2$, $\eta(<1)$ is a phenomenological next gain factor for the ASE mode, ϵ is the spontaneous emission coefficient, Ψ is the external phase angle, and q is the effective coupling coefficient between the ASE mode E_a and compound-cavity lasing eigenmodes E_{sm}.

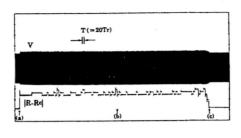

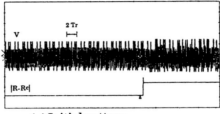

(a) Initial pattern

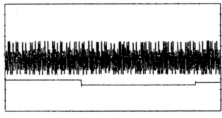

(b) Chaotic pattern

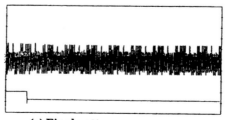

(c) Final pattern

Figure 1.34: (a) Chaotic search time series of output V and error signal $|R - R_c|$. (b) Expanded chaotic search time series for points marked with (a), (b) and (c) in (a).

Supplement 2: Model Equation of Hybrid Bistable Device with Delay

Theoretically, the optoelectronic hybrid device with a time delay is modelled by the following equation:

$$T_m dV(t)/dt = -V(t) + V_0 + \mu F(V(t - t_R)), \qquad (S.5)$$

where V is the electrical voltage applied to the E-O modulator, T_m is the response time of the feedback, t_R is the delay time, μ is the effective loop gain proportional to the input optical power, F is the E-O characteristic of the modulator, and V_0 is the off-set voltage.

[Textbooks and Reviews]

[A] K. Shimoda, T. Yajima, Y. Ueda, T. Shimizu and T. Kasuya, *Quantum Electronics*, Shokabo Publishing, Tokyo 1972 (in Japanese).

[B] N. B. Abraham, P. Mandel, and L. M. Narducci, *Dynamical Instabilities and Pulsations in Lasers*, Progress in Optics XXV (Elsevier Science Publisher, 1988).

[C] K. Otsuka, "Nonlinear phenomena in semiconductor lasers", *Proceedings of SPIE on Nonlinear Optics and Materials*, Vol. 1497 (1991) pp. 432–443.

[D] K. Ikeda, "Delay-differential equations modelling nonlinear optical resonators", in *Optical Instabilities*, Eds. by R. W. Boyd, M. G. Raymer and L. M. Narducci (Cambridge University Press, New York, 1986) pp. 85–98.

[E] Paul Mandel, *Theoretical Problems in Cavity Nonlinear Optics*, Cambridge University Press, 1997, Chapter 6.

References

1) E. Lorenz, "Deterministic non-periodic flow", *J. Atmos. Sci.*, Vol. 20 (1963) pp. 130–141.

2) H. Haken, "Analogy between higher instabilities in fluids and lasers", *Phys. Lett. A*, Vol. 53 (1975) pp. 77–78; "Generalized Ginzburg-Landau equations for phase transition-like phenomena in lasers, nonlinear optics, hydrodynamics and chemical reactions", *Z. Phys. B*, Vol. 21 (1975) pp. 105–114.

3) H. Zeghache, P. Mandel, N. B. Abraham, and C. O. Weiss, "Phase and amplitude dynamics in the laser-Lorenz model", *Phys. Rev. A*, Vol. 38 (1988) pp. 3128–3131.

4) C. O. Weiss, N. B. Abraham, and U. Hubner, "Homoclinic and heteroclinic chaos in a single-mode laser", *Phys. Rev. Lett.*, Vol. 61 (1988) pp. 1587–1590.

5) H. Zeghache and P. Mandel, "Influence of detuning on the properties of laser equations", *J. Opt. Soc. Am.*, Vol. B2 (1985) pp. 18–22.

6) C. O. Weiss, "Chaotic laser dynamics", *Opt. Quantum Electron.*, Vol. 20 (1988) pp. 1–22.

7) a) L. W. Casperson, "Spontaneous coherent pulsations in laser oscillators", *IEEE J. Quantum Electron.*, Vol. QE-14 (1978) pp. 756–761; b) L. W. Casperson, "Spontaneous coherent pulsations in ring-laser oscillators", *J. Opt. Aoc. Am.*, Vol. B2 (1985) pp. 62–72; c) L. W. Casperson, in *Laser Physics*, Lecture Notes in Physics, Vol. 182, Ed. J. D. Harvey and D. F. Walls (Springer, Heidelberg) p. 88.

8) S. T. Hendow and M. Sargent III, "The role of population pulsation in single-mode laser instabilities", *Opt. Commun.*, Vol. 40 (1982) pp. 385–390; "Theory of single-mode laser instabilities", *J. Opt. Soc. Am.*, Vol. B2 (1985) pp. 84–101.

9) P. Mandel, "Effect of Doppler broadening on the stability of monomode ring lasers", *Opt. Commun.*, Vol. 44 (1983) pp. 400–404.

10) R. S. Gioggia and N. B. Abraham, "Routes to chaotic output from a single-mode, dc-excited laser", *Phys. Rev. Lett.*, Vol. 51 (1980) pp. 650–653.

11) F. T. Arecchi, R. Meucci, G. P. Puccioni, and J. R. Tredicce, "Experimental evidence of subharmonic bifurcation, multistability, and turbulence in a Q-switched gas laser", *Phys. Rev. Lett.*, Vol. 49 (1982) pp. 1217–1220; T. Midavaine, D. Dangoisse, and P. Glorieux, "Observation of chaos in a frequency-modulated CO_2 laser", *Phys. Rev. Lett.*, Vol. 55 (1985) pp. 1989–1992.

12) W. Klische, H. R. Telle and C. O. Weiss, "Chaos in a solid-state laser with a periodically modulated pump", *Opt. Lett.*, Vol. 9 (1984) pp. 561–563.

13) T. Mukai, T. Saitoh, K. Otsuka, and K. Kubodera, "Observation of FM-laser oscillations including a subharmonic resonance effect in an external cavity semiconductor laser", *International Quantum Electronics Conference*, San Francisco, 1986, paper MCC3 (unpublished).

14) S. E. Harris and O. P. McDuff, "Theory of FM laser oscillation", *IEEE J. Quantum Electron.*, Vol. QE-1 (1965) pp. 245–262.

15) D. Baums, W. Elsasser, and E. Göbel, "Farey tree and devil's staircase of a modulated external-cavity semiconductor laser", *Phys. Rev. Lett.*, Vol. 63 (1989) pp. 155–158.

16) R. Lang and K. Kobayashi, "External optical feedback effects on semiconductor injection laser properties", *IEEE J. Quantum Electron.*, Vol. QE-16 (1980) pp. 347–355.

17) J. Mørk, J. Mark, and B. Tromborg, "Route to chaos and competition between relaxation oscillations for a semiconductor laser with optical feedback", *Phys. Rev. Lett.*, Vol. 65 (1990) pp. 1999–2002.

18) A. P. Bogatov, P. G. Eliseev, P. G. Ivanov, and B. N. Sverdlov, "Anomalous interaction of spectral modes in a semiconductor laser", *IEEE J. Quantum Electron.*, Vol. QE-11 (1975) pp. 510–515.

19) T. Mukai and T. Saitoh, "Detuning characteristics and conversion efficiency of nearly degenerate four-wave mixing in a 1.5 μm traveling-wave semiconductor laser amplifier", *IEEE J. Quantum Electron.*, Vol. QE-26 (1990) pp. 865–875.

20) T. Mukai and K. Otsuka, "New route to optical chaos: successive-subharmonic-oscillation cascade in a semiconductor laser coupled to an external cavity", *Phys. Rev. Lett.*, Vol. 55 (1985) pp. 1711–1714.

21) J.-D. Park, D.-S. Seo, and J. G. McInerney, "Self-pulsations in strongly coupled asymmetric external cavity semiconductor lasers", *IEEE J. Quantum Electron.*, Vol. QE-26 (1990) pp. 1353–1362.

22) K. Otsuka and T. Mukai, "Asymmetrical coupling, locking and chaos in a compound cavity semiconductor laser", *Proceedings of SPIE on Chaos*, Vol. 667, (1986) pp. 122–129.

23) See, for example, K. Peterman, "External optical feedback phenomena in semiconductor lasers", *IEEE J. Selec. Topics Quantum Electron.*, Vol. 1 (1995) pp. 480–489 and references therein.

24) J. Sacher, W. Elsasser, and E. Göbel, "Intermittency in the coherence collapse of a semiconductor laser with external feedback", *Phys. Rev. Lett.*, Vol. 63 (1989) pp. 2224–2227.

REFERENCES

25) a) I. Fisher, G. H. M. van Tartwijk, A. M. Levine, W. Elsasser, E. Göbel, and D. Lenstra, "Fast pulsing and chaotic itinerancy with a drift in the coherence collapse of semiconductor lasers", *Phys. Rev. Lett.*, Vol. 76 (1996) pp. 220–223; (b) G. Vaschenko, M. Giudici, J. J. Rocca, C. S. Menoni, J. R. Tredicce, and S. Balle, "Temporal dynamics of semiconductor lasers with optical feedback", *Phys. Rev. Lett.*, Vol. 81 (1998) pp. 5536–5539.

26) Y. Silberberg and I. Bar-Joseph, "Optical instabilities in a nonlinear Kerr medium", *J. Opt. Soc. Am.*, Vol. B1 (1984) pp. 662–670.

27) H. M. Gibbs, F. A. Hopf, D. L. Kaplan, and R. L. Shoemaker, "Observation of chaos in optical bistability", *Phys. Rev. Lett.*, Vol. 46 (1981) pp. 474–477.

28) H. Nakatsuka, S. Asaka, H. Itoh, K. Ikeda, and M. Matsuoka, "Observation of bifurcation to chaos in an all-optical bistable system", *Phys. Rev. Lett.*, Vol. 50 (1983) pp. 109–112.

29) R. G. Harrison, W. J. Firth, and I. A. Al-Saidi, "Observation of bifurcation to chaos in an all-optical Fabry-Perot resonator", *Phys. Rev. Lett.*, Vol. 53 (1984) pp. 258–261.

30) K. Otsuka and H. Iwamura, "Analysis of a multistable semiconductor light amplifier", *IEEE J. Quantum Electron.*, Vol. QE-19 (1983) pp. 1184–1186; K. Otsuka and H. Iwamura, "Theory of optical mustistability and chaos in a resonant-type semiconductor laser amplifier", *Phys. Rev. A*, Vol. 28 (1983) pp. 3153–3155.

31) H. Kawaguchi, "Multiple bistability and multistability in a Fabry-Perot laser diode amplifiers", *IEEE J. Quantum Electron.*, Vol. QE-23 (1987) pp. 1429–1433.

32) K. Otsuka and H. Kawaguchi, "Period-doubling bifurcation in detuned lasers with injected signals", *Phys. Rev. A*, Vol. 29 (1984) pp. 2953–2956.

33) H. Kawaguchi, K. Inoue, T. Matsuoka, and K. Otsuka, "Bistable output characteristics in semiconductor laser injection locking", *IEEE J. Quantum Electron.*, Vol. QE-21 (1985) pp. 1314–1317.

34) K. Ikeda and K. Matsumoto, "High-dimensional chaotic behavior in systems with time-delayed feedback", *Physica D*, Vol. 29 (1987) pp. 223–235.

35) T. Aida and P. Davis, "Applicability of bifurcation to chaos: Experimental demonstration of methods for switching among multistable modes in a nonlinear resonator", in *OSA Proc. Nonlinear Dynamics in Optical Systems*, N. B. Abraham, E. M. Garmire, and P. Mandel, Eds., Vol. 7 (1991) pp. 540–544.

36) T. Aida and P. Davis, "Oscillation mode selection using bifurcation of chaotic mode transitions in a nonlinear ring resonator", *IEEE J. Quantum Electron.*, Vol. QE-30 (1994) pp. 2986–2997.

Chapter 2

REFERENCE MODELS OF LASER COMPLEX SYSTEMS

In Chapter **1**, nonlinear dynamics and chaotic behaviors in prototypical optical systems *with small degrees of freedom*, including Lorenz-Haken Rabi precession instabilities in homogeneously broadened single-mode lasers, mode-splitting Casperson instabilities in inhomegeneously broadened single-mode lasers, modulation and light-feedback induced instabilities in semiconductor laser diodes featuring relaxation oscillations as well as longitudinal mode spacing oscillations, and delayed-feedback induced Ikeda instabilities in nonlinear passive optical resonators, are described.

In the following chapters, discussion will shift towards nonlinear dynamics and chaotic behaviors in optical "complex systems" with large coupled degrees of freedom. Some peculiar dynamical properties of such complex systems are described. In this Chapter 2, several 'reference models' of laser complex systems are introduced and their fundamental properties are described. Here, the 'reference model' implies that it is an idealized laser model from which the generic features of systems with large degrees of freedom can be extracted, although it is not so easy for practical implementations. They include homogeneously-broadened multimode class-C laser systems in section **2.1** [A], modulated globally-coupled multimode class-B laser systems with equal modal gain in section **2.2** [B], and globally-coupled multimode intracavity second-harmonic generation class-B lasers with equal modal gain but in different parameter regimes in section **2.3** [C] and in section **2.4** [D].

The first model (section **2.1**), in which the homogeneous linewidth can support multiple cavity modes, has yet to be realized in practice at present; however, quantum-wire and quantum-dot semiconductor lasers, in which a strong oscillator strength (e.g., increased $\gamma_\parallel$ and decreased $\gamma_\perp$) is expected resulting from a quantum confinement effect, might have a possibility to become a good candidate for realistic devices. The latter two globally-coupled multimode laser systems (sections **2.2**–**2.4**) can be implemented, in principe, in the form of globally-coupled laser arrays consisting of identical laser elements.

2.1 Maxwell-Bloch Turbulence in Homogeneously-Broadened Multimode Laser Model [A]

Assume lasers, in which the homogeneous linewidth $\gamma_\perp$ or the inhomogeneous line width γ^* is sufficiently larger than the cavity mode spacing frequency $f_c = c/2L$, many cavity modes whose resonance frequencies are within the gain band have the potential for contributing to laser oscillation. Let us call these kind of lasers multimode lasers. This section will discuss typical instability and chaos which occur in these kinds of laser systems. Although in realistic lasers, the effect of inhomogenous broadening is important, since the contribution of this effect is much too complex, the discussion of this section will not take the effect of inhomogeneous broadening into consideration. Therefore, we will consider homogeneously broadened multimode lasers.

2.1.1 Resonant Rabi instability

First, the electric field and material field are expanded as

$$E(z,t) = \sum_q \tilde{E}(q,t)U_q(z)e^{-iqct}, \qquad (2.1a)$$

$$\rho(z,\omega,t) = \sum_q \tilde{\rho}(\omega,q,t)U_q(z)e^{-iqct}, \qquad (2.1b)$$

$$\Delta\rho(z,\omega,t) = \sum_q \tilde{\Delta\rho}(\omega,q,t)U_q, \qquad (2.1c)$$

where all variables are expanded in the complete orthogonal system, i.e., cavity modes

$$U_q(z) = \exp(iqz),$$

where the wavenumber is given by $q = 2n\pi/L$ (n: integer). Here, a ring resonator is assumed with the periodic boundary condition of $E(t, z=0) = E(t, z=L)$.

Maxwell-Bloch equations for multimode laser oscillations are transformed into the following coupled-mode equations

$$d\tilde{E}(q,t)/dt = -\kappa\tilde{E}(q,t) - (N\mu\omega/2\epsilon)\tilde{\rho}(\omega,q,t), \qquad (2.2)$$

2.1. MAXWELL-BLOCH TURBULENCE

$$d\tilde{\rho}(\omega,q,t)/dt = [i(cq+\omega-\omega_0)-\gamma_\perp]\tilde{\rho}(\omega,q,t)$$
$$-(2\pi\mu/h)\sum_{q_1+q_2=q}\tilde{E}(q_1,t)\tilde{\delta\rho}(\omega,q_2,t)e^{iq_2ct}, \quad (2.3)$$

$$d\tilde{\Delta}\rho(\omega,q,t)/dt = \gamma_\|[\Delta\rho_0 - \tilde{\Delta}\rho(\omega,q,t)] + (\pi\mu/h)$$
$$\cdot \sum_{q_1+q_2=q}[\tilde{E}(q_1,t)\tilde{\rho}^*(\omega,-q_2,t)e^{-i(q_1+q_2)ct} + c.c.]. \quad (2.4)$$

Similarly to the case of single-mode lasers, let us introduce the electric field variable F

$$F(t,z) = -\mu E(t,z),$$

where $|F|$ denotes the Rabi frequency. To deal with the issue of multimode lasers, the multimode Maxwell-Bloch equations (2.2)–(2.4) are used. Unlike the single mode laser systems, the stationary solution of multimode lasers is not unique. Suppose that only a single mode q_0 is excited, we immediately obtain the following family of solutions from Eqs. (2.2)–(2.4).

$$F_{qs} = \frac{\delta_{q,q_0}F_s\sqrt{1-c^2q_0^2/\omega^{*2}}\exp(-icq_0t)}{\gamma_\perp + \kappa}, \quad (2.5a)$$

$$\tilde{\rho}_{qs} = \frac{\kappa}{s^2}[1+i\frac{cq_0}{\gamma_\perp+\kappa}]F_{qs}, \quad (2.5b)$$

$$\tilde{\Delta}\rho_{qs} = \frac{\kappa\gamma_\perp}{s^2}[1+\frac{c^2q_0^2}{(\gamma_\perp+\kappa)^2}]\delta_{q,0}. \quad (2.5c)$$

Here, F_s is the stationary amplitude of resonant mode $q_0 = 0$

$$F_s = [\gamma_\|\gamma_\perp(\frac{\alpha_0}{\alpha_{th}^{(1)}}-1)]^{1/2}, \quad (2.6)$$

where ω^* is given by

$$\omega^* = (\gamma_\perp+\kappa)F_s/\sqrt{\gamma_\perp\gamma_\|} \quad (2.7)$$

and this quantity approximately equals the power broadened width $\sqrt{\gamma_\perp/\gamma_\|}F_s$ which is given in Eq. (1.9). We call this family of solutions the *single mode stationary solution* (SSS). Therefore, the characteristic of multimode lasers is the existence of many SSS, which oscillate in different frequencies.

From the above equation, it appears that all modes satisfying $|q| \leq \omega^*/c$ are able to oscillate. However, gains of modes for $|cq| \gg \gamma_\perp$ are small and these modes cannot oscillate in competition with resonant modes with $|cq| \ll \gamma_\perp$.

We discuss the stability of SSS confining the analysis to the resonant solution $q_0 = 0$. Let us return to the modal equations (2.2)–(2.4) and obtain three characteristic values λ_q of the Jacobian matrix $\mathbf{G}_q$ of the following linearized equation related to small deviations $\delta F_q, \delta \rho_q, \delta w_q$

$$\frac{d\delta \mathbf{x}_q}{dt} = \mathbf{G_q} \delta \mathbf{x_q}. \tag{2.8}$$

Here, $\delta \mathbf{x}_q$ is the column vector consisting of $\delta F_q, \delta \rho_q, \delta w_q$ of mode q. The problem of instability in multimode lasers was first considered by Risken and Nummendal [1] and also by Graham and Haken [2].

Rather than describing details of the linear stability analysis, let us intuitively describe the essential difference between instabilities of multi-mode and single-mode lasers through a discussion which is similiar to that given in **1.3**. As in **1.3**, assume that $\gamma_\perp = \gamma_\parallel \equiv \gamma$ and the amplitude of the mode with wavenumber q is sufficiently small. In this case, only the coupling terms of mode 0 and mode $\pm q$ are effective in the equation of motion for mode q. Therefore, the equation of motion for ρ_q, for instance, is approximated as

$$d\tilde{\rho}_q/dt = -\gamma \tilde{\rho}_q + F_q \Delta \rho_0 + \tilde{\Delta}\rho_q F_0. \tag{2.9}$$

from Eq. (2.3). If this equation is expressed in terms of the following quantities,

$$Z_q^{(\pm)} \equiv s^2[(\tilde{\rho}_q + \tilde{\rho}^*_{-q})/2 \pm i(\tilde{\Delta}\rho_q + \tilde{\Delta}\rho^*_{-q})/2], \tag{2.10}$$

$$V_q \equiv (F_q + F^*_{-q})/2 \tag{2.11}$$

the following equations are obtained from Eqs. (2.2)–(2.4):

$$dV_q/dt = (-iqc - \kappa)V_q + \frac{1}{2}(Z_q^{(+)} + Z_q^{(-)}) \tag{2.12a}$$

$$dZ_q^{(\pm)}/dt = -\gamma Z_q^{\pm} \pm iZ_0^{(\pm)} V_q \pm iF_0 Z_q^{(\pm)} \tag{2.12b}$$

Let us compare Eq. (2.12) with the linearized equation (1.22) of the single-mode laser Lorenz system. Making the following transformations of variables, $(V_q, Z_q^{(+)}, Z_q^{(-)}) \rightarrow (\delta F, \delta Z, \delta Z^*)$, $(F_0, Z_0^{(+)}, Z_0^{(-)}) \rightarrow (F_s, Z_s^{(+)}, Z_s^{(-)})$ it is clear that the former equation looks very similar to the latter. However, the difference exists in Eq. (2.12), in which the electric field oscillates at the characteristic frequency cq. In fact, this term dramatically affects the characteristics of instability of the system.

2.1. MAXWELL-BLOCH TURBULENCE

In the limit of large pumping power as has been done in **1.2.1**, $Z_q^\pm$ begins a Rabi precession at frequency F_0 since the third-term becomes predominant in Eq. (2.12) and hence we may set $Z_q^\pm(t) = \exp(\mp iF_0 t)\hat{Z}_q^{(\pm)}(t)$, where $\hat{Z}_q^\pm(t)$ are slowly varying parts. Consequently, the electric field V_q also oscillates at the same frequency and V_q is given by

$$V_q = \frac{1}{2} \frac{Z_q^{(\pm)}}{i(cq \mp F_0) + \kappa}. \tag{2.13}$$

This corresponds to Eq. (1.23). If this equation is substituted into Eq. (2.12b), it is easily understood that the renormalized damping constant in the strong pumping limit becomes

$$\gamma \to \gamma - \frac{F_0}{2} \frac{(F_0 \mp cq)/\kappa}{[(F_0 \mp cq)^2/\kappa^2 + 1]} \tag{2.14}$$

for the $Z_q^\pm$ mode. Here, $Z_0 \to s^2 F_0/k$ is used for the strong pump limit. Consequently, the modes nearly resonant with the Rabi precession satisfying the following conditions

$$|(F_0 \mp cq)/\kappa| < O(1), \tag{2.15}$$

becomes unstable, yielding self-oscillation of Rabi precession when the stationary amplitude $F_0 = F_s$ (Eq. (2.6a)) becomes large enough. Different from single-mode lasers, this instability occurs whether the cavity is bad or good. The reason is that among many cavity modes, the modes which are able to be resonant with the Rabi precession always exist. Therefore, it is appropriate to call this instability *resonant Rabi instablity*. On the contrary, Lorenz instability which occurs in single-mode lasers shall be called non-resonant Rabi instability.

According to the rigorous linear stability analysis, the threshold value $\alpha_{th}^{(2)}$ is found to be unrelated to κ and it is given as follows:

$$\alpha_{th}^{(2)}/\alpha_{th}^{(1)} = 4 + 3(\gamma_\parallel/\gamma_\perp) + 2\sqrt{4 + 6(\gamma_\parallel/\gamma_\perp) + 2(\gamma_\parallel/\gamma_\perp)^2}. \tag{2.16}$$

If α_0 (linear gain) exceeds the second threshold $\alpha_{th}^{(2)}$, the "unstable Rabi bands" of cq appear around $RC_\pm = \pm cq_R$ and broadens with increase in the pump as shown in Fig. 2.1, in which a good cavity condition is assumed. Modes within these band will break into oscillation. It is not so difficult to show that the frequency Im $\lambda_{q=q_R}$ of the Rabi unstable mode is close to the Rabi frequency near the bifurcation point.

Just above the second threshold, only one mode appears in each unstable band. In this domain, the resonant Rabi instability, featuring a

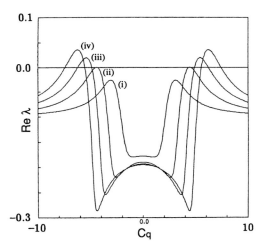

Figure 2.1: Real part of stability exponent $Re\lambda$ as a function of characteristic frequency cq for various values of linear gain α_0. α_0 is increased in the order (i)→(ii)→(iii)→(iv).

simple sinusoidal oscillation, takes place. With a slight increase in the pump, the number of modes in unstable bands increases and the system makes a transition to a periodic Rabi oscillation whose envelope is modulated at the cavity mode spacing frequency v_g/L as shown in Fig. 2.2(a), where $v_g = (d\,\mathrm{Im}\,\lambda_q/dq)_{q=q_R}$ is the group velocity. This implies that the relative phase of Rabi oscillation among modes in the unstable bands is self-organized to be fixed to a unique value. The spatio-temporal pattern of electric field is given by

$$f(t,z) = \exp i(q_c x - \mathrm{Im}\,\lambda_{q_c} t) f(x - v_g t) + F_s, \qquad (2.17)$$

where $f(x - v_g t)$ is the envelope function $f(x) = \sum_n A_n \exp i(2n\pi/L)x$ with the period L. A_n is the complex amplitude of the n-th mode and has a significant value only when its wavenumber q_n is within the unstable band.

However, such a beautiful self-mode locking type oscillation state exists only in a narrow pump power regime after the resonant Rabi oscillation. If the pump power is increased above the self-mode-locking regime, the phase-locking fails and chaotic motions develop in modes within the Rabi unstable bands and the output waveform begins to go into disarray.

2.1.2 Self-induced mode partition instability

Figure 2.3 shows a rough sketch of the phase diagram of the dynamical states obtained in the present model, assuming a good cavity condition. A

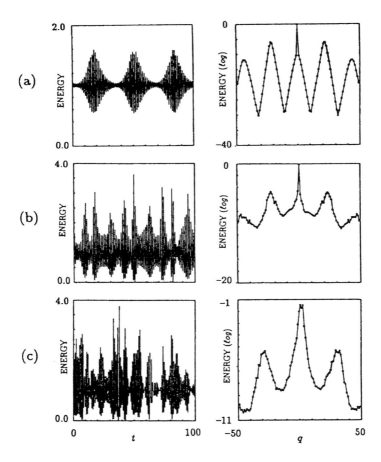

Figure 2.2: Waveforms (left) and energy spectra (right) of typical dynamical states above the first threshold. (a) self-locked state, (b) local chaos, (c) global chaos. Here, $\kappa = 0.1$, $\gamma_\perp = \gamma_\parallel = 1$ and $L = 2\pi$.

transition to chaos always occurs at the higher side of pumping level, and there are two types of chaotic solutions. We show in Figs. 2.2(b)(c) typical examples of the modal energy spectrum together with the waveforms for the two chaotic solutions, namely *local chaos* and *global chaos*. In the good cavity case, the peaks in the spectrum locate at the gain center ($q = 0$), at the Rabi unstable wavenumbers ($q = q_R$) and their higher harmonics. The main feature of chaos at the lower pump level (b) is that most of the energy is monopolized by the mode at the gain center ($q = 0$), whereas the energy is distributed over all the modes in the gain band $|cq| < \gamma_\perp$ for the chaotic solution appearing at the higher pump level (c).

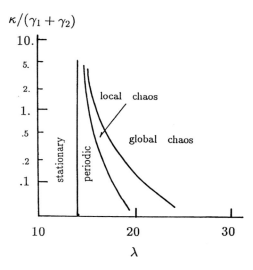

Figure 2.3: Phase diagram of dynamical states.

To understand the difference between the two chaotic solutions, we have to consider a simple topological feature inherent in the solutions of the Maxwell-Bloch equations. Let us first consider what happens when the spatial coupling is discarded by setting $c = 0$ in the original Maxwell partial differential equation with respect to t and z. Then the M-B equation describes an assembly of spatially *uncoupled* single mode laser rate equation, each of which obeys the laser Lorenz equation (1.10)–(1.12):

$$d\mathbf{x}_z(t)/dt = \mathbf{F}(\mathbf{x}_z(t)), \qquad (2.18)$$

where $\mathbf{x}_z(t) = (F(t,z), \rho(t,z), w(t,z))^T$. In the good cavity limit, the solution relaxes toward the stationary lasing solution such that $L^\pm = (F_s e^{i\phi}, \rho_s e^{i\phi}, w_s)$. An important fact is that the stationary solution is allowed to have an arbitrary phase. This is because the M-B equation is invariant for the phase shift transformation $(F, \rho) = (F, \rho)e^{-i\phi}$. Thus the set of variables describing the uncoupled M-B system relaxes onto the cylinder

$$|F(z)| = F_s \qquad (0 \le z \le L) \qquad (2.19)$$

defined in the phase space. Because of the periodic boundary condition, the cylinder may be identified as the torus T. As long as the system obeys the spatially uncoupled dynamics, the phase on the torus is determined by the initial condition.

In the actual system, however, the spatial coupling will make the spatial configuration of ϕ relax toward a fixed stationary configuration on a

2.1. MAXWELL-BLOCH TURBULENCE

long time scale related with the spatial coupling strength. The single mode stationary solution (SSS) described by Eq. (2.5) represents such fixed configurations. Thus there are two steps for establishing SSS: The first step is a rapid relaxation onto the torus T, and the second one is a slow reorientation process of phase on T toward a stable spatial configuration. The slowness of the second process reflects the fact that the spatial coupling causing relaxation on T is quite weak. We can easily change the spatial pattern of phase continuously with all $\mathbf{x}_z(t)$ sticking to T, whereas it is not easy to take $\mathbf{x}_z$ off from the torus T.

From these observations we can conclude that there exists an invariant quantity; that is the winding number associated with the spatial pattern of phase

$$W = (\phi(L) - \phi(0))/2\pi \qquad (2.20)$$

We emphasize again that to change W on T without introducing any spatial discontinuous structure is impossible. It is easy to see that the SSS with the mode q_0 being excited is a stable spatial pattern with a given winding number $W = q_0/(2\pi/L)$. Since the motion on T is quite slow, we may call the subspace corresponding to T in the phase space the *slow manifold*. There are a number of slow manifolds with different winding numbers, but the stable slow manifolds are those with winding numbers less than

$$|W| < \gamma_\perp/c \times (2\pi/L) = t_R \gamma_\perp \qquad (2.21)$$

as is anticipated from the stable condition for the SSS (Eq. (2.5)).

Having the above arguments in mind, let us consider chaotic motions far beyond the threshold of the resonant Rabi instability. If we neglect the influence of the chaotic modes around the Rabi unstable band, then the system would relax toward a stable SSS on a slow manifold according to the mechanism mentioned above. In actual situations, however, the chaotic motion excited at the two Rabi unstable bands $\delta_1 q_R$ and $\delta_2 q_R$ around q_R and $-q_R$ yields a fluctuating force of small wavenumbers $q = (q_R \pm \delta_1 q_R) + (-q_R \pm \delta_2 q_R) \sim 0$, which are located in the gain center region, which is referred to as the *core region* C hereafter. Thus such a chaotic force, namely the Rabi force, excites fluctuation on the torus T, which is composed of the modes in the resonant gain band, i.e., in the region C. Being trapped on T, the total field amplitude is equal to F_s hence $F(t,z) = F_s \exp[i\phi(t,z)]$, while the spatial phase pattern $\phi(t,z)$ running with the light velocity is perturbed by the Rabi force and fluctuates very slowly in time. Then the amplitude of mode q is given by

$$F_q = F_s \int_0^L e^{i(\phi(t,z)-qz)} dz. \qquad (2.22)$$

With the stationary phase approximation, it is easy to see that F_q has a significant magnitude at wavenumbers other than q_0 if there exist a (z,t) such that $d\phi(z,t)/dz = q$. The appearance of the spatial nonuniformity in the phase derivative (i.e., phase inflection) implies that *modes which have wavenumbers without well-defined winding number* are excited together with the resonant lasing mode q_0.

Not far above the second threshold, the amplitude of the Rabi force is small, and the energy is monopolized almost entirely by the resonant lasing mode $q_0 = W/(2\pi/L)$. With an increase in the pumping level, the amplitude of the Rabi force is enhanced, and the rate of energy destributed over the modes other than q_0 in the core region increases according to Eq. (2.22). Thus, although the total energy $\int_0^L |F(t,z)|^2 dz = |F_s|^2 L$ is almost conserved, energy partitioned to modes other than q_0 fluctuates violently on a quite long time scale. Temporal evolution of modal outputs and their summation for modes in the core region are shown in Fig. 2.4. Such a fluctuation phenomenon has often been observed in the multimode oscillation of semiconductor lasers and it has been called as *mode partition noise*.

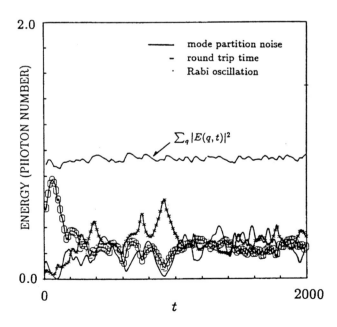

Figure 2.4: Temporal evolution of modal outputs and their summation for modes in the central core region. $\alpha_0 = 30$ and other parameters are the same as Fig. 2.2. Fluctuation time scales are given by the length of bars.

It should be noted that in semiconductor lasers mode partition noise appears only through spontaneous emission, while in the present system it is brought about by the self-induced chaotic Rabi force created in the Rabi unstable bands. Indeed, to examine the effect of the Rabi chaotic bands, we observe what occurs when one of the two Rabi bands is 'removed' from the M-B system. The removal is achieved by making the amplitudes of all the modes in one of the two Rabi bands zero at each step of numerical integration. At this moment, chaotic variation of the core mode ceases after the removal operation and a single mode stationary solution (SSS) is eventually realized. This domain of self-induced mode partition instability corresponds to *local chaos* in Fig. 2.3.

2.1.3 Self-induced mode hopping: chaotic itinerancy

Well above the second threshold, the mode partitioning becomes more active and the modal distribution of energy gets broadened around $q = q_0$ as shown in Fig. 2.2(c). Then a drastic phenomenon takes place; the winding number, which has been supposed to be stable and invariant, temporally changes on a time scale much longer than that of the mode partition noise. The sudden change of the winding number is detected in the form of a change of the mode number q_0 into which most of the energy is partitioned. Such a phenomenon corresponds to the mode-hopping which has been often observed in a variety of laser systems. Temporal change of the winding number means that paths connecting the slow maifolds, which have been isolated in the domain of local chaos, are formed dynamically in the phase space. In this way the isolated local attractors merge with each other to form an unified global attractor.

Let me show in Fig. 2.5(a) a direct evidence showing the formation of a path. When a transition takes place between two manifolds, a spatially localized object with a loop structure is nucleated in the spatial pattern of the complex amplitude of the electric field. It is easy to understand that the winding number increases by 1 through forming such an object temporally (Fig. 2.5(b)). A typical example of temporal evolution of photon numbers of modes in the core region C and the change of winding number are shown in Fig. 2.6. The time scale is taken so long that the mode partition fluctuation looks like a rapid variation. A remarkable fact is that on such a long time scale the distribution of photon numbers over the core (central) modes often changes drastically. In this figure, the number of the most predominantly excited mode changes successively as $n = 0 \to n = 1 \to n = 2 \to \cdots$.

If the force acting on the slow manifold (i.e., central core modes) is an externally applied noise, it will not be easy to nucleate such a highly structured object. In our system, the force acting on the slow manifold is due to the chaotic motion excited in the Rabi unstable bands, and chaos

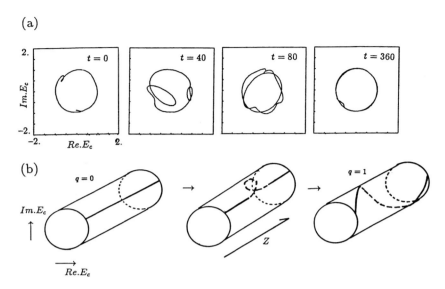

Figure 2.5: (a) Temporal evolutions of the spatial trajectory of the complex field in the central core region in the global chaos regime, (b) spatial configuration of the complex field corresponding to the single stationary solutions (SSS) with $q = 0$ and $q = 1$, and typical configuration appearing in the transition from the $q = 0$ to $q = 1$ solution, in which a 'loop structure' is nucleated.

induced there possesses a strong correlation with the dynamical state in the slow manifold. In other words, the Rabi force *knows* the state of the slow manifold, and if a small loop is formed in the slow manifold the Rabi force is deformed so as to enhance the loop structure. Thus chaos enables the system to wander over different local attractors more efficiently than the externally applied noise.

Indeed, to verify the interplay between the core region and the unstable Rabi bands, random variables which are decorrelated from the state of modes in the core region are 'transplanted' to the Rabi chaotic bands. At this moment, clear transition between well-defined winding numbers like Fig. 2.6 cannot be realized if the random variables are constructed so as to replicate the statistical property of the modes in the Rabi unstable bands. Therefore, the Rabi force is not a simple random force generated independently of the state of the system, but it is an 'internal force' which is formed with a strong correlation with the past evolution of the core region. Such a self-induced switching among the ruins of local attractors is called *chaotic itinerancy** [see Footnote 1], in which an external random force is useless for switching. This phenomenon is generic in complex systems and often appears in this book hereafter.

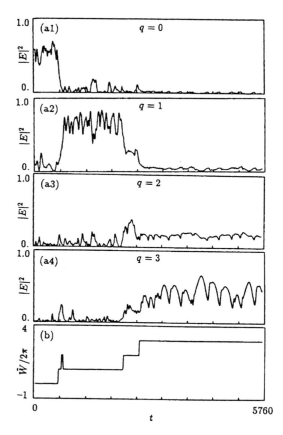

Figure 2.6: Temporal evolution of photon numbers (a) and winding number (b) of the central core modes exhibiting successive mode hopping.

2.1.4 Physical interpretation and information theoretic analysis

Self-induced mode partition noise and mode-hopping phenomena described in the previous subsection can be interpreted in terms of parametric four-wave mixing process among modes located in the central core region C and two Rabi unstable bands $RC_\pm$, which are described in the previous Chapter **1.4, 1.5**. In other terms, due to the third-order nonlinear susceptibility $\chi^{(3)}$ in the laser medium, the scattering process like

$$RC_+ + RC_- \to C, \quad R_+ + C \to RC_-, \quad etc. \qquad (2.23)$$

occurs through three-wave interaction. As a result, interplay between central core modes and chaotic Rabi band modes takes place, leading to mode partition noise and mode hopping depending on the strength of Rabi force created in the unstable bands $RC_\pm$.

On the other hand, quantitative information theoretic analyses based on Shannon's mutual information have been demonstrated [A]. Here, only the 3-point mutual information map involving C and $RC_\pm$ bands is presented (For further details, see [A]). An example of 3-point mutual information maps is shown in Fig. 2.7, in which RC_+ at $q_s = 45$ serves as a 'sender'. One peak indicated by $\rightarrow$ is situated at the 'receiver' q_r in the C region and at the "hidden" ("additional") sender $q_{s'}$ in RC_-, respectively, and vice versa. This map clearly indicates the existence of mutual information flows among C and $RC_\pm$ bands corresponding to Eq. (2.23) in this system.

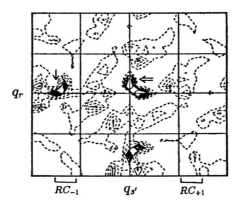

Figure 2.7: 3-point mutual information map among core modes and unstable Rabi bands.

2.2 Modulation Dynamics of Globally Coupled Multimode Lasers [B]

In this section, we describe modulation dynamics in the reference model of globally-coupled multimode class B lasers with spatial hole-burning focussing on inherent antiphase dynamics and related phenomena. Different from the previous section of multimode class-C laser models, due to the lack of the degree of freedom resulting from the adiabatic elimination of polarization dynamics, additional degree of freedom is necessary to raise nonlinear dynamics leading to chaos. In the present model, a pump modulation is employed as an additional degree of freedom. This is the simplest way to examine nonlinear behaviors in class-B lasers as was mentioned in **1.4**.

2.2.1 Introduction

Antiphase periodic state (abbreviated as APS) discovered by Hadley and Beasley in coupled Josephson junctions [3] is a property displayed by systems in which N (> 1) degrees of freedom oscillate with a strong phase correlation. This type of dynamics is now well established for the self-pulsing state in coupled chemical oscillators [4] and spatio-temporal electroenecephalogram (EEG) pattern generation in olfactory sensing systems [5]. In laser systems, Kubodera and Otsuka demostrated antiphase periodic oscillations in a laser-diode-pumped microchip two-mode solid-state laser for the first time [6]. In that state, all oscillators are in the same periodic state, but each oscillator is shifted by $1/N$ of a period from its neighbor [3]. This implies a strong correlation among the phases of modal intensities. There coexist $(N-1)!$ antiphase periodic states in such globally-coupled oscillator systems. Therefore, attractor crowding [7] occurs and the basin of attraction shrinks rapidly as the system size N increases.

In this section, let me describe dynamical properties of APS in the model of deeply modulated Tang-Statz-deMars's globally-coupled multimode lasers, which incoporate cross-saturation of population inversion among modes due to the spatial hole-burning effect [8] on the basis of intensitive numerical analysis for wide parameter regions, focusing on the destabilization process of APS's when the number of oscillating modes increases. Section **2.2.2** summarizes antiphase periodic spiking states in deeply modulated multimode lasers and phase diagrams of APS for the simplest case of $N = 3$, where only two equivalent APS's coexist, are presented. When N increases, the basin of attraction of APS's shrinks; APS attractors are kicked out from the main bifurcation branch and APS tends to coexist with periodic states which have a larger basin of attraction. In such regimes, however, APS's are selectively excited from the periodic states by applying seeding light-pulses. In section **2.2.3**, seeding-induced assignment to a desired APS and the required condition for seeding light-pulse are examined. As the number of modes increases further, APS's, which coexisted with periodic states, are destabilized and grouping into periodic (chaotic) antiphase spiking modes to periodic (chaotic) sustained relaxation oscillation modes takes place. In this regime, *grouping chaos*, which features cooperative synchronization of chaotic sustained relaxation oscillation modes, appears. Section **2.2.4** treats this grouping phenomenon which appears in the destabilizing process of APS's. Finally, applicability of antiphase periodic states to factorial dynamic memory is discussed in section **2.2.5**

2.2.2 Equations of motion and antiphase periodic states

A. Modulated Tang-Statz-deMars multimode laser model

For the globally-coupled lasers, we employ homogeneously broadened class-B lasers with spatial hole-burning, in which polarization dynamics can be adiabatically eliminated. The following equations are derived from Maxwell-Bloch laser equations by the adiabatic elimination of polarization:

$$dn_0/dt = w_0 + \Delta w_m \cos(\tau\omega_m t) - n_0 - \sum_{i=1}^{N} \gamma_i(n_0 - n_i/2)s_i, \qquad (2.24)$$

$$dn_k/dt = \gamma_k n_0 s_k - n_k(1 + \sum_{i=1}^{N} \gamma_i s_i), \qquad (2.25)$$

$$ds_k/dt = K[\{\gamma_k(n_0 - n_k/2) - 1\}s_k + \epsilon n_0 + s_{i,k}], \quad k = 1, 2, \ldots, N. \quad (2.26)$$

Here, $t = T/\tau$ is the normalized time (τ is the population lifetime), w_0 is the pump power normalized by the first lasing mode threshold, Δw_m is the modulation amplitude, ω_m is the modulation frequency, n_0 is the constant term of the spatial Fourier expansion of the population inversion density normalized by the first lasing mode threshold, n_k is the first-order Fourier component of population inversion density for the k-th mode, s_k is the normalized photon density, γ_k is the gain ratio to the first lasing mode, $K = \tau/\tau_p$ (τ_p is the photon lifetime), ϵ is the spontaneous emission coefficient, and $s_{i,k}$ is the injection-seeding signal to the k-th mode. Note that each lasing mode globally couples with all other modes through cross saturation of population inversion. In the followings, let us assume uniform gain distribution among modes, i.e., $\gamma_k = 1$. These equations are easily modified to include globally coupled laser arrays having a uniform gain [9]. (For a detailed deviation of Eqs. (2.24)–(2.26) for class-B lasers with spatial hole-burning, see textbook [T]).

The linear stability analysis reveals that an N-mode free-running laser is always stable in time, and the relaxation oscillation at $\omega_r = [(w-1)/\tau\tau_p]^{1/2}$ is damped out. If the modulation amplitude increases to where the pump power drops below the threshold during part of the pump modulation cycle, the total output behaves just like a single-mode laser and exhibits spiking-mode oscillations in a wide modulation frequency region $\omega_m = \omega_s < \omega_r$, while each mode exhibits N-alternative spiking pulsations at ω_s/N [B-1]. [For *spiking-mode oscillation**, see Footnote 2]. That is the antiphase periodic state, resulting from the cross-saturation mechanism of population inversions. In short, there coexist $(N-1)!$ antiphase periodic states in the phase space. As for $N = 1$, the lowest repetition frequency edge of the spiking mode (optimum spiking mode frequency), at which the

2.2. MODULATION DYNAMIC

highest peak power is obtained, is given analytically [6], [11]

$$\omega_{s,opt} = \pi[2ln(2\epsilon\sqrt{K})^{-1}]^{-1/2} \times [(w_0 - 1)/(w_m - 1)]^{1/2}\omega_r, \qquad (2.27)$$

where $w_m = w_0 + (2/\pi)\Delta w_m$ denotes the pulse height of an equivalent rectangular pump pulse.

B. Bifurcation diagram for the simplest case of $N = 3$

Bifurcation diagrams for the simplest case of $N = 3$, where only two equivalent APS's coexist, are shown in Fig. 2.8 as a function of modulation frequency, assuming different pump power w_0 and modulation amplitude Δw_m. Here, SRO, CS, CRO denote in-phase sustained relaxation oscillation, clustered state, and chaotic relaxation oscillations. A clustered state is a state in which modes are divided into different periodic oscillations. Stable antiphase periodic states are realized in the region denoted by APS. The optimum spiking mode frequency for the single mode case $\omega_{s,opt}(N = 1)$, which is given by Eq. (1), is shown by arrows. From Fig. 2.8, $\omega_{s,opt}(N = 3)$ (e.g., the lowest frequency edge of APS) is found to shift to a higher frequency compared with $\omega_{s,opt}(N = 1)$. This results from the cross-saturation of population inversions by different modes. Examples of different types of oscillations, such as SRO, CS, APS and CRO, are shown in Fig. 2.9.

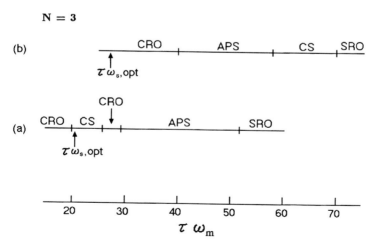

Figure 2.8: Bifurcation diagram for $N = 3$. $K = 10^3$, $\epsilon_k = 1.2 \times 10^{-7}$. (a) $w_0 = 2.7$, $\Delta w_m = 2.0$, (b) $w_0 = 4.0$, $\Delta w_m = 3.3$. SRO: in-phase sustained relaxation oscillation; CS: clustered state; CRO: chaotic relaxation oscillation; APS: antiphase periodic state.

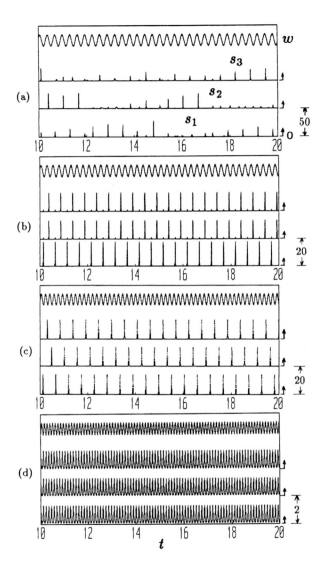

Figure 2.9: Typical oscillation waveforms of SRO, CS, APS and CRO at different modulation frequencies. (a) CRO at modulation frequency $\Omega_m \equiv \tau\omega_m = 20$, (b) CS at $\Omega_m = 25$, (c) APS at $\Omega_m = 35$, (d) SRO at $\Omega_m = 55$. Other parameters are the same as Fig. 2.8(a).

2.2.3 Seeding-induced excitation of APS's

A. Coexistence of periodic motions and APS's

When N increases, the basin of attraction of the antiphase states shrinks very rapidly: antiphase attractors are kicked out from the main bifurcation diagram and tend to coexist with the periodic states such as SRO and CS, forming an isolated branch. In general, these periodic states (e.g., SRO and CS) have a larger basin of attraction than do the antiphase states, and the system is almost always attracted by these periodic orbits when starting from arbitrary initial conditions. However, one can assign the system from these periodic states to antiphase states embedded in the high-dimensional phase space by injecting small light pulses to $(N-1)$ modes in the desired sequences as seeds at time intervals of $2\pi/\omega_s$, only during the $(N-1)$ modulation cycle [B-1], [12]. Figure 2.10(a) is an example of antiphase periodic states realized by injection seeding signals in the sequence of $\{1,2,3,4,5\}$ starting from the periodic clustered state $[(s_1,s_2,s_3),(s_4,s_5)]$, assuming $N=5$, $w_0=2.7$, $\Delta w_m=2.0$, $\tau\omega_m=35$, $K=1000$, seed-pulse height $s_{i,k}=0.2$, seed-pulse width $=0.06$, and $\epsilon=1.2\times 10^{-7}$. It is very likely that the injection seeds apply a driving force to the system such that the trajectory falls on the basin of attraction of the isolated antiphase state starting from the period-2 cycle clustered state. Phase space trajectories on the $[g_i,s_i]$ plane ($i=1$) is depicted in Fig. 2.10(b), where $g_i=n_0-n_i/2$ is proportional to a modal gain. The

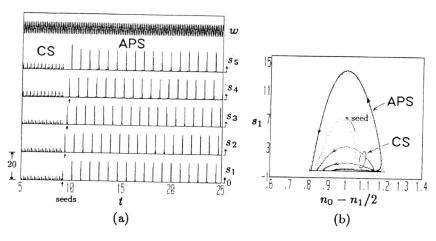

Figure 2.10: Assignment to the antiphase periodic state by injection seeding for a five-mode laser. The laser is initially in the clustered state $[(s_1,s_2,s_3),(s_4,s_5)]$, in which the modes are separated into two groups. The adopted parameters are presented in the text. (a) Intensity waveform, (b) Phase space trajectory for the $k=1$ mode.

switching from the CS to the APS is clearly seen. If the seeding condition, which will be discussed in C, is not satisfied, the system returns to the periodic state (e.g., clustered state in this case) after staying around the neighborhood of the APS attractor. This point will be discussed in section **2.2.5**.

B. Bifurcation diagrams and circulation analysis

Bifurcation diagrams for four- and five-modes are shown in Figs. 2.11 and 2.12 for different w_0 and Δw_m. Results are shown for with and without injection seeding pulses. By changing the order of sequence of applied seeding pulses, one can easily assign the system to a desired APS state. Here, let us carry out the proposed circulation analysis to examine self-organization nature in multimode lasers [12]. Figure 2.13 shows mode-to-mode gain circulations $G_{i,j}$ and the summation over the closed circulation path $\sum G_{i,j}$:

$$\sum G_{i,j} = G_{1,2} + G_{2,3} + G_{3,4} + G_{4,5} + G_{5,1}.$$

It is seen that the vanishing gain circulation rule, i.e., $\sum G_{i,j} \ll G_{i,j}$, which implies the dynamical equipartition of population inversion among modes [12], is established for APS. Here,

$$G_{i,j} = g_i \dot{g}_j - \dot{g}_i g_j, \qquad (2.28)$$

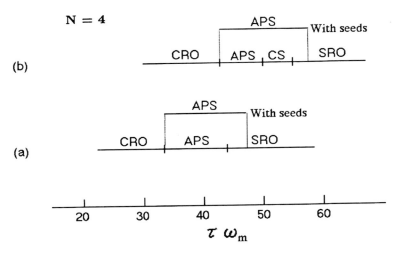

Figure 2.11: Bifurcation diagram for $N = 4$ with and without seeds. Adopted parameter values are same as those of Figs. 2.8(a) and 2.8(b), respectively. seed-pulse height $s_{i,k} = 0.2$ and pulsewidth $\Delta t = 0.06$. Without seeds, the system returns to the main branch outside APS region.

2.2. MODULATION DYNAMIC

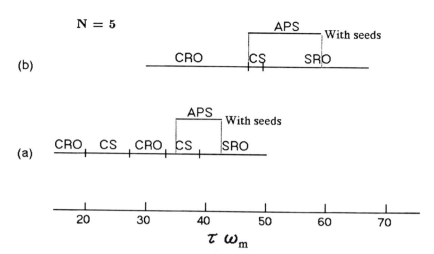

Figure 2.12: Bifurcation diagram with and without seeds for $N = 5$. Adopted parameter values are the same as those of Fig. 2.8(a) and 2.8(b), respectively, and seed-pulses are same as Fig. 2.11.

where the overdot indicates the time derivative. When $G_{i,j} > 0$ (< 0), the gain flow occurs from $i(j)$-th mode to $j(i)$-th mode. This vanishing gain circulation rule is confirmed to hold for *arbitrary* closed gain circulation path involving $3 \leq n \leq N$ different modes [12]. (To be specific, $\sum G_{i,j} = 0$ for periodic states like SRO and CS, while the $\sum G_{i,j}$ value is negligibly small compared with $G_{i,j}$ but non-zero for APS and CRO.) The gain circulation analysis will appear again in the next Chapter **3**.

C. Inverse power-law relation for seeding assignment to APS's

As is shown in **2.2.3 A**, with seeding pulses, one can switch the system from the periodic state to desired APS patterns. Here, we investigate the seed-pulse intensity $s_{i,k}$ required for a successful switching as a function of seed-pulse width Δt. A relation between the pulse intensity and pulse width which is required for the system to switch from the periodic state to the APS is shown in Fig. 2.14, where the adopted parameter values are the same as Fig. 2.10. The successful switching is realized in the shadowed region. It is interesting to note that the lower and upper bounds of the seed-pulse intensity are expressed by the inverse power-law relation, i.e., $s_{i,k} \propto \Delta t^{-1}$. The lower bound, in particular, provides the threshold energy for successful switching. If the seeding pulse energy is too high, the strong depletion of population inversion occurs and switching fails. This results in the appearance of the upper bound.

If one introduces a Newtonian mechanical picture, the inverse-power

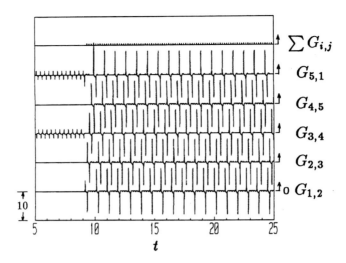

Figure 2.13: Gain circulation analysis of APS for $N = 5$, which corresponds to Fig. 2.10.

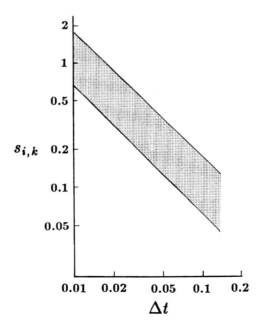

Figure 2.14: Relation between the seed-pulse height and pulsewidth which is required for the system to switch from the periodic state to the APS for $N = 5$. Adopted parameter values are the same as Fig. 2.10.

2.2. MODULATION DYNAMIC

law suggests that the motion along the switching path takes place according to $\dot{D} = Cs_{i,k}$ (driving force), where D is the distance from the periodic attractor in the phase space and C is constant. The lower bound provides the distance between the periodic attractor and the APS attractor. The switching 'band structure' in Fig. 2.14 may correspond to the basin of attraction of the APS attractor. The highest frequency bound nearly corresponds to the period of APS, e.g., $2\pi/\tau\omega_m$, beyond which there is no sense in seeding.

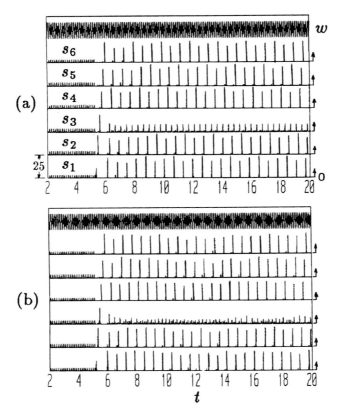

Figure 2.15: (a) Excitation of the $(5, 1)$ grouping periodic state by injection seeding for a six-mode laser. (b) $(5, 1)$ grouping chaos state. Adopted parameter values are given in the text.

2.2.4 Grouping phenomenon in a large number of modes

A. Grouping into APS's and sustained relaxation oscillations

As N increases further up to six-modes, the antiphase periodic attractors on an isolated branch can no more exist and are replaced by grouping states. This results from the fact that the spiking period of APS becomes shorter than the recovery time of population inversion of each mode as N increases and all modes cannot form APS. Grouping states are excited from the coexisting periodic states by the injection seeding similarly to **2.2.3**. In this grouping, one subset of the lasing modes exhibits periodic (chaotic) spiking antiphase oscillations, while the other subset shows

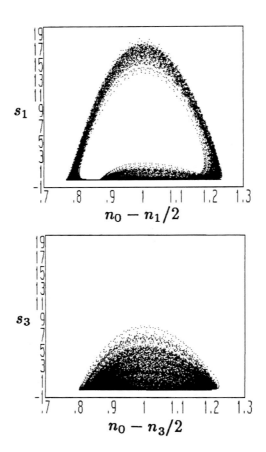

Figure 2.16: Phase space trajectory corresponding to the $(5, 1)$ grouping chaos state shown in Fig. 2.15(b).

2.2. MODULATION DYNAMIC

periodic (chaotic) small amplitude sustained relaxation oscillation. Without seeding, on the other hand, the system always relaxes to the periodic states, such as SRO and CS, starting from arbitrary initial conditions. Figure 2.15(a) shows the result for $N = 6$, $w_0 = 4.2$, $\Delta w_m = 3.3$, and $\tau w_m = 48.55$, where the system is in the SRO state initially and injection seeding switches the system to the grouping periodic state, in which five modes exhibit antiphase periodic spiking oscillation while one mode shows small amplitude periodic sustained relaxation oscillation. To be specific, the APS mode group exhibits period-3 cycle solution.

Figure 2.15(b) shows the result when τw_m was increased to 50, where other parameters are the same as (a). In this case, five modes exhibit chaotic antiphase spiking oscillations, in which the APS phase relationship is nearly maintained among modes, while one mode shows small amplitude chaotic sustained relaxation oscillation. Phase space trajectories corresponding to Fig. 2.15(b) are depicted in Fig. 2.16 except for transients. A Lyapunov spectrum for Fig. 2.15(b) is calculated to be (1.26, 0, −0.38, −1.07, −1.89, −2.45, −3.28, −5.19, −5.39, −7.16, −11.40, −13.11, −17.45, −21.69), where one positive Lyapunov exponent exists in this case.

A bifurcation diagram with and without seeding is depicted in Fig. 2.17(b). It is interesting to point out that the vanishing gain circulation rule, along arbitrary closed gain circulation path involving $3 \leq n \leq N$ different modes, is also established in such $(5, 1)$ grouping states similarly to APS shown in Fig. 2.13. (For instance, even for the fundamental case

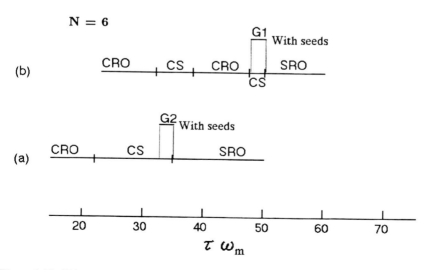

Figure 2.17: Bifurcation diagram with and without seeds for $N = 6$. Adopted parameter values are given in the text.

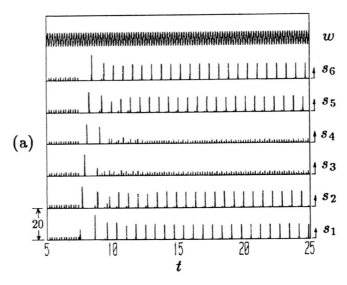

Figure 2.18: (a) Excitation of the (4, 2) grouping quasi-periodic state by injection seeding for a six-mode laser. (b) (4, 2) grouping chaos state. (c) Enlargement of (b) for synchronized chaotic modes. See the text for parameters.

involving three modes including one chaotic relaxation oscillation mode, this rule is approved.)

B. *Grouping chaos and cooperative synchronization*

When the pump power is decreased below $w_0 = 3.3$, four out of six modes exhibit antiphase periodic (chaotic) spiking oscillations while other two modes show synchronized periodic (chaotic) sustained relaxation oscillations. It is easily shown that there coexist $N!(N-3)!/2!(N-2)! = 90$ equivalent grouping chaos states in the phase space and that one can assign the system to a desired grouping state by changing the sequence of seeding pulses. Examples of such periodic and chaotic grouping oscillations are shown in Figs. 2.18(a) and (b), where $w_0 = 2.7$, $\Delta w_m = 2.0$, $\tau \omega_m = 34.45$ (a) and $\tau \omega_m = 35$ (b) are assumed. Other parameters are the same as Fig. 2.10. In the case of periodic grouping shown in Fig. 2.18(a), both mode groups exhibit quasi-periodic motions, in which two modes show synchronization. Enlargement of Fig. 2.18(b) for synchronized chaotic modes is shown in Fig. 2.18(c). A phase space trajectory for grouping chaos state (Fig. 2.18(b)) is depicted in Fig. 2.19 after omitting transients. A bifurcation diagram with and without seedings are shown in Fig. 2.17(a). An example of return maps of peak intensities for chaotic APS modes and synchronized sustained relaxation oscillation modes are depicted in Fig. 2.20,

2.2. MODULATION DYNAMIC 79

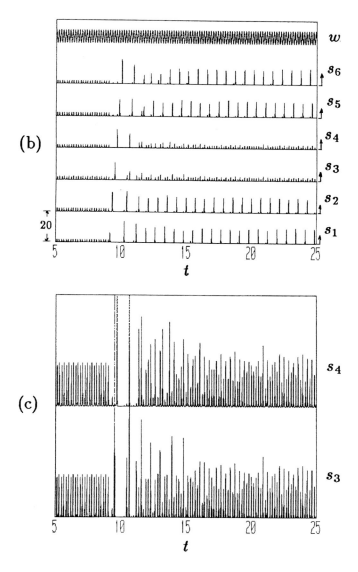

Figure 2.18: (continued).

except for transients, where the mode 1 and 4 are synchronized in this example. A chaotic evolution of each mode and perfect synchronization between the two modes (s_1, s_2) are clearly seen in the figure. A calculated Lyapunov spectrum was (0.46, 0.05, 0, −0.19, −0.43, −1.06, −1.85,

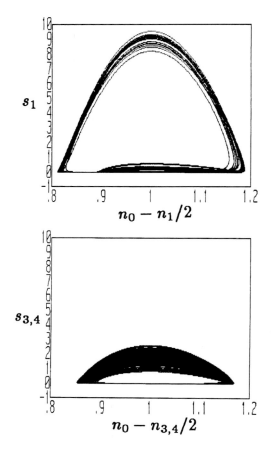

Figure 2.19: Phase space trajectories for chaotic APS mode ($k = 1$) and synchronized chaos modes ($k = 3, 4$).

-2.29, -2.36, -3.58, -11.43, -13.59, -14.27, -21.16), where two positive Lyapunov exponents exist.

Let us investigate the cross-saturation dynamics that lead to grouping chaos based on the circulation analysis. Such a characterization of synchronized evolution by a dynamical correlation function would effectively explain time-dependent interplays between lasing modes. (In the present globally-coupled system, it appears impossible to separate the elements (e.g., lasing modes) of a chaotic laser into master and slave subsystems on the contrary to the case of unidirectional interaction [13].)

Figure 2.21 shows $G_{i,j}(t)$ corresponding to Fig. 2.18(b), where the sum of gain circulations along the closed gain circulation path $\sum G_{i,j}$ is also

2.2. MODULATION DYNAMIC

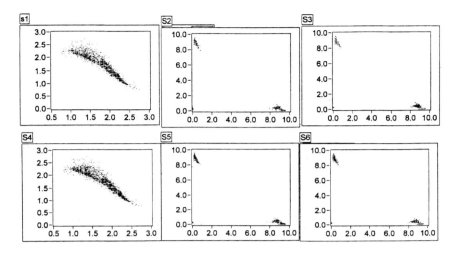

Figure 2.20: Return map indicating synchronized chaos in the $(4, 2)$ grouping chaos state, where the $k = 1$ and 4 modes are synchronized in this case.

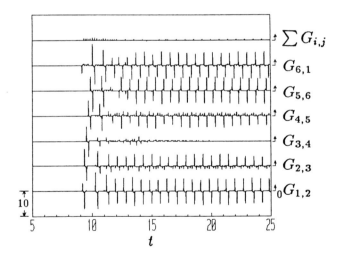

Figure 2.21: Characterization of the $(4, 2)$ grouping chaos shown in Fig. 2.18(b) by gain circulations.

plotted. Before applying seeds, $G_{i,j} = 0$ because all the modes are in the in-phase SRO state, i.e., $s_i(t) = s_j(t)$. From this figure, it is clearly seen that the gain circulation $G_{3,4}$ between the potentially synchronizing modes, $k = 3$ and 4, tends to approach zero as a result of the injection seeding after some transients, while other gain circulations oscillate in time. Another interesting feature is that this kind of $(4, 2)$ grouping is realized, satisfying the vanishing gain circulation rule of $\sum G_{i,j} \ll G_{i,j}$ as well.

Therefore, *the vanishing gain circulation rule acts as the strong constraint for all kinds of dynamical states, including SRO, CS, CRO, APS and grouping states, in the present system which governs self-organized nature of nonlinear dynamical behaviors.* This rule physically implies the dynamical equipartition of population inversions among modes [12], which manifests itself in generic antiphase dynamics observed experimentally in multimode lasers which will be described in Chapter **3** again.

Let us consider the connection between the vanishing gain circulation rule and the synchronization for the $(4, 2)$ grouping. Assume the fundamental gain circulation path involving three modes including one antiphase spiking mode s_r and two potentially synchronizing modes s_p and s_q. An easy way to satisfy the vanishing gain circulation rule is found to be $s_p(t) = s_q(t)$. This is nothing more than synchronization. In this case, the rule is automatically satisfied because $G_{p,q} = 0$ and $G_{r,p} = -G_{q,r}$, e.g., $G_{r,p} + G_{p,q} + G_{q,r} = 0$. (Another possible antiphase motion for s_p and s_q, which ensures this rule shown in Fig. 2.13, is forbidden in this case.) In short, the vanishing gain circulation rule strongly requires the synchronization.

To understand the effect of injection seeding and to characterize the interplay between the antiphase modes and the synchronized chaotic modes from another point of view, let us examine the intensity circulation [12]:

$$S_{i,j} = s_i \dot{s}_j - \dot{s}_i s_j. \qquad (2.29)$$

Figure 2.22 shows intensity circulations between the synchronized chaotic mode group (s_3, s_4) and the antiphase mode group, i.e., $\sum S_{3,j}$ and $\sum S_{4,j}$ $(j = 1, 2, 5, 6)$, and their difference. It is interesting to note that *unidirectional* chaotic intensity flows from the antiphase mode group $s_a \equiv \sum s_j (j = 1, 2, 5, 6)$ to the synchronized mode group (s_3, s_4) are brought about by injection seeding (i.e., $\sum S_{3,j} < 0$ and $\sum S_{4,j} < 0$). This implies that antiphase modes are excited by injection seeding (see Fig. 2.18(b)) and the unidirectional power flow, which perturbs a subset of modes s_3 and s_4, is created cooperatively. As a result, the lasing modes s_3 and s_4 tend to be perturbed by the common chaotic signal s_a as time evolves, leading to synchronized chaos $s_3(t) = s_4(t)$. The same unidirectional power flow from the APS mode group to synchronized periodic relaxation oscillation modes occurs in the case of Fig. 2.18(a) as well. Such a nonreciprocal independence of intensity (i.e., energy) flow will be discussed again in **2.4**.

A phase diagram for obtaining grouping states $(5, 1)$ and $(4, 2)$ on the $[w_0, \tau w_m]$ plane is summarized shown in Fig. 2.23, where G1 denotes the

2.2. MODULATION DYNAMIC

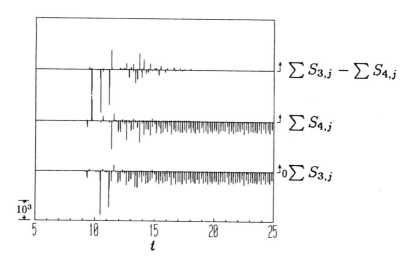

Figure 2.22: Characterization of the $(4,2)$ grouping chaos shown in Fig. 2.18(a) by intensity circulations.

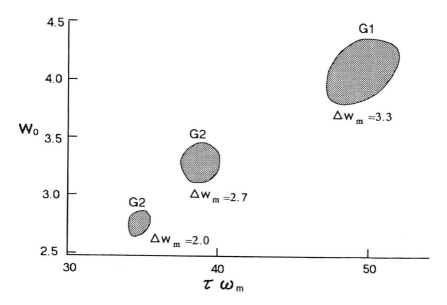

Figure 2.23: Phase diagram for obtaining grouping state $(5,1)$ and $(4,2)$ on the $[w_0, \tau\omega_m]$ plane, where G1 denotes $(5,1)$ state and G2 does $(4,2)$ state.

former type of grouping (5, 1), G2 does the latter type of grouping (4, 2). Here, the grouping into periodic mode groups such as Figs. 2.15(a) and 2.18(a) occurs in the lower frequency region of each zone and grouping chaos develops as the modulation frequency increases within each zone.

Similar numerical experiments to Fig. 2.14 were carried out to identify required condition for seeding pulses to realize a successful switching from periodic states to the grouping state. It is found that the similar inverse power-law relation holds in this case, however, the successful switching region (e.g., switching 'band structure') shrinks greatly as compared with Fig. 2.14. This implies that the basin of attraction of grouping states is extremely narrow, although it is finite.

When N is greater than 6, such grouping states were found to disappear in entire parameter regions in the present system; only periodic states, such as SRO and CS, and the chaotic state CRO are present in the phase space. If the spontaneous emission coefficient ϵ is decreased, the basin of attraction of APS's tends to increase [9].

2.2.5 Factorial dynamic pattern memory

A direct assignment to desired dynamical APS patterns by a 'systematic' small seeding-pulse injection suggests the possibility of factorial dynamic pattern memories. In this section we discuss two points: 1) How to erase APS patterns? 2) How about the connectivity between different APS patterns?

A. Stagnant motion around APS attractors

The most relevant way to erase APS patterns and kick the system back to the periodic patterns is the application of 'uniform' pulses to all the modes simultaneously. To check this idea, let us first assign the system to the APS pattern and then apply the erase-pulses to all the modes simultaneously. Typical example of such simulations are shown in Fig. 2.24 for different erase-pulse height, where pulsewidth $\Delta t = 0.06$ and other adopted parameter values are the same as Fig. 2.10. In the case of (a), the system returns to the periodic state rapidly. In the case of (b), on the other hand, the system stays around the neighborhood of APS attractor for a long time, whose time scale is much longer than other lasing time scale, and finally the system returns to the periodic state. It is interesting to note that such "stagnant motion" features chaotic switching among the ruins of chaotic grouping states described in **2.2.4** *A* (e.g., Fig. 2.15(b)), in which one small-amplitude chaotic relaxation oscillation mode switches chaotically as indicated by arrows, while other modes form chaotic APS state. This stagnant motion features 'transient' chaotic itinerancy discussed in **2.1** terminated by a clustered state.

2.2. MODULATION DYNAMIC

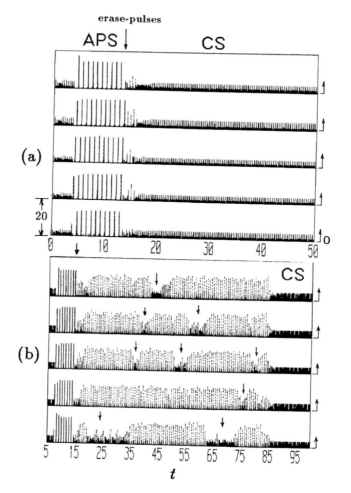

Figure 2.24: Switching from the APS to the clustered state by simultaneous application of uniform erase-pulses, assuming $\Delta t = 0.06$. Adopted parameters are the same as Fig. 2.10. (a) $s_{i,k} = 0.1$, (b) $s_{i,k} = 0.2$.

Intensive numerical simulations for various erase-pulse shapes shows that such stagnation motions occur more frequently than a rapid return to the periodic state like (a). Therefore, the erase procedure is not reliable in terms of dynamic pattern memory applications since the system is frequently attracted by such a stagnation layer presumably existing in basin boundaries on the switching path created by 'uniform' injection pulses. In the case of grouping states G1 and G2 discussed in **2.2.4**, on the other hand, the system is found to return to the periodic states more rapidly by 'uniform' erase-pulses, suggesting a simpler stucture around basin boundaries of grouping states.

B. Flexibility of re-writable dynamic pattern memories

Now, let us examine the flexibility of direct re-writing of APS dynamic patterns by successive applications of different seed-pulse patterns. In this study, let us carry out switching simulations for all possible combinations for $N = 4$ for brevity. An example of switching between different APS patterns is shown in Fig. 2.25. Table 2.1 summaries the lowest re-writing pulse height required for quick switching between APS patterns, assuming the pulsewidth $\Delta t = 0.06$. This table indicates the flexibility of re-writing memories. From this table, it is found that flexible symmetrical switchings for all possible combinations are ensured in this case. Furthermore, it is interesting to note that the switching path is not uniform and there exist some very-easy switching paths for particular pairs. This may result from the difference in short-term population dynamics triggered by seeding pulses near the switching point for different pairs. The switching path formation is investigated by changing the seed-pulse shapes similarly to Fig. 2.14 and the similar inverse power-law relation for the threshold pulse

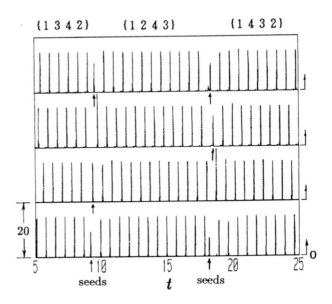

Figure 2.25: Switching among different APS patterns by successive applications of different seeding pulses, assuming $N = 4$, $K = 1000$, $w_0 = 2.7$, $\Delta w_m = 2.0$, $\tau w_m = 35$, $\epsilon = 1.2 \times 10^{-7}$, $s_{i,k} = 0.2$ and $\Delta t = 0.06$.

2.2. MODULATION DYNAMIC

Table 2.1: Switching diagram among different APS patterns for $N = 4$ showing the lowest switching intensities, assuming $\Delta t = 0.06$.

	1 2 3 4	1 2 4 3	1 3 4 2	1 3 2 4	1 4 3 2	1 4 2 3
1 2 3 4		0.27	0.09	0.13	0.25	0.26
1 2 4 3	0.27		0.24	0.26	0.003	0.14
1 3 4 2	0.09	0.24		0.20	0.003	0.27
1 3 2 4	0.13	0.26	0.20		0.002	0.20
1 4 3 2	0.25	0.003	0.003	0.002		0.28
1 4 2 3	0.26	0.14	0.27	0.20	0.28	

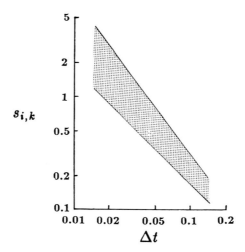

Figure 2.26: Relation between seed-pulse height and pulsewidth which is required for the system to switch between different APS attractors for $N = 4$.

energy is found. An example for the switching from $\{1, 2, 3, 4\}$ to $\{1, 2, 4, 3\}$ is shown in Fig. 2.26. It is interesting to note that the upper bound obeys a relation $s_{i,k} \propto \Delta t^{-3/2}$.

2.3 Chaotic Itinerancy in Antiphase Intracavity Second-Harmonic Generation (ISHG) [C]

The APS in the multimode ISHG system was first observed by Baer in two-mode lasers [14] and then by K. Wiesenfeld *et al.* in dual-polarization multimode regimes [15]. In this section, the emergence of local chaotic antiphase states and self-induced switching among the ruins of local chaotic antiphase attractors (i.e., chaotic itinerancy) leading to fully developed global chaos is presented by numerical simulations in a model of intracavity second-harmonic generation (ISHG) in multimode lasers, in which autonomous pulsations due to the second-order nonlinear process among lasing modes take place through a Hopf bifurcation.

Dynamical characterization of chaotic itinerancy is carried out by the circulation analysis.

2.3.1 Introduction

Chaotic itinerancy (abbreviated as CI), which implies a self-induced switching among the ruins of chaotic attractors occurring generally in the transition process from local to global chaos, is a generic phenomenon in nonlinear systems with large degrees of freedom and was described in this chapter so far.

Such coexisting antiphase periodic states are expected to be destabilized when the control parameter is changed. Our motivation of the present study was to find what kind of phenomenon is brought about when these equivalent APS's are destabilized. In this section, numerical results are shown in a model of a multimode laser with intracavity second harmonic generation (abbreviated as ISHG) [14] in which the elements $\{e_m\}$ of the system are the active modes of the laser. The self-induced switching among the ruins of local chaotic APS attractors (i.e., chaotic itinerancy) is found to take place in the transition process from the local chaotic APS to fully developed chaos. The physical process occurring in switching points is shown by the intensity circulation analysis, which was discussed in the previous section **2.2**.

2.3.2 Multimode ISHG equations and Hopf bifurcation

The multimode laser problem involving intracavity second-harmonic generation can be described by the following set of equations [14].

$$\dot{I}_k = K(G_k - \alpha - g\epsilon I_k - 2\epsilon \sum_{j \neq k} \mu_{jk} I_j) I_k, \qquad (2.30)$$

2.3. CHAOTIC ITINERANCY IN ANTIPHASE

$$\dot{G}_k = \gamma - (1 + I_k + \beta \sum_{j \neq k} I_j)G_k, \qquad k = 1, 2, \ldots, N \tag{2.31}$$

Here, the overdot means derivation with respect to time t, $t = T/\tau_f$ is the normalized time (τ_f is the population lifetime), I_k is the intensity for the k-th longitudinal mode, G_k is the modal gain, $K = \tau_f/\tau_c$ (τ_c is the cavity round-trip time), α is the cavity loss, γ is the small-signal gain and β is the cross-saturation parameter. The term ϵ is the nonlinear coupling coefficient of doubling crystals and g is a geometrical factor whose value depends on the orientation of the laser crystal relative to the doubling crystal as well as the phase delays due to their birefringence. The lasing modes are specified to oscillate in either of two orthogonal polarizations. In Eq. (2.30), $\mu_{jk} = g$ if j and k have the same polarization, while $\mu_{jk} = 1 - g$ if the two modes have orthogonal polarizations. We have made the simplifying approximation that α, γ, β and ϵ are the same for all modes. For small values of the linear gain γ, the solutions of Eqs. (2.30)–(2.31) are stable steady states. Above a critical value of the linear gain, the steady state is destabilized and periodic solutions emerge. For even larger gains, the time-dependent solutions display the full variety of complex behaviours, including chaos.

Wang and Mandel determined the general criterion of stability for stable steady-state multimode operation in Eqs. (2.30)–(2.31) [16, T]. They analytically constructed the emerging periodic solutions for when that condition is not satisfied and showed that they display antiphase periodic states (APS). According to their analysis, the APS appears via a Hopf bifurcation when the small-signal gain γ exceeds the following critical value in the case of single polarization oscillation $[M, P] = [N, 0]$ in the limit of $\epsilon \to 0$, $1/K \to 0$, and α, β, γ, g and $K\epsilon = O(1)$, where M is the number of modes polarized along one direction and P is the number of modes oscillating in the orthogonal polarization.

$$\gamma_H = \frac{\alpha}{1 - [1 + (N-1)\beta]/(gK\epsilon)} + O(1/K) \tag{2.32}$$

provided the geometrical factor that verifies the inequantity

$$g > [1 + (N-1)\beta]/K\epsilon. \tag{2.33}$$

In the following analysis, we study the case of single polarization for brevity.

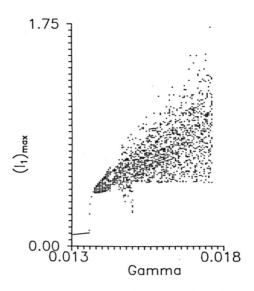

Figure 2.27: Bifurcation diagram for five-mode, single polarization oscillation. Adopted parameter values are $K = 500000$, $\alpha = 0.01$, $g = 0.52$, $\epsilon = 0.00005$ and $\beta = 0.6$.

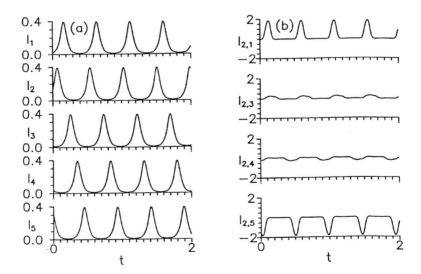

Figure 2.28: Antiphase periodic state (APS), assuming $\gamma = 0.01365$. Other parameters are the same as in Fig. 2.27. (a) Temporal evolutions of modal intensities. (b) Corresponding intensity circulations from the k = 2 mode to all other modes. (c) Phase portrait on the $[I_k, G_k]$ plane and state points of different modes at a particular time. t is the time.

2.3.3 Bifurcation process and circulation analysis

A. Antiphase periodic state (APS)

Figure 2.27 shows the bifurcation diagram in the case where five modes oscillate with the same polarization, i.e., $[M, P] = [5, 0]$. Here, successive maxima of $I_1(t)$ are plotted as a function of the small-signal gain γ. Adopted parameter values are chosen according to the experiment [15].

Figure 2.28(a) shows an example of APS that appeared above a Hopf bifurcation point $\gamma_H = 0.0135$, where a stable APS in the order of $\{2, 1, 3, 4, 5\}$ is clearly seen.

Here, we employ the intensity circulation defined as [12]

$$I_{i,j} = I_i \dot{I}_j - \dot{I}_i I_j. \tag{2.34}$$

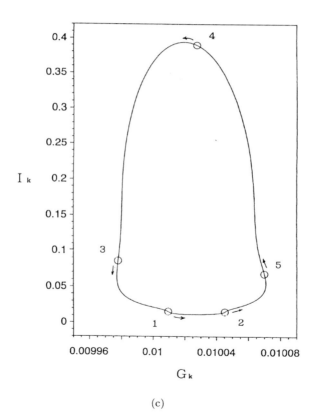

(c)

Figure 2.28: (continued).

Here, I_k is an equivalent value to s_k (photon density) in (2.29). $I_{i,j} > 0$ (< 0) implies an intensity flow from mode i (j) to mode j (i). The intensity circulations $I_{2,j}$ ($j = 1, 3, 4, 5$) between the mode (i.e., $k = 2$) and all other modes, corresponding to Fig. 2.28(a), are shown in Fig. 2.28(b). From this figure, we see that intensity (i.e., power) flow occurs from the $k = 2$ mode to the next mode $k = 1$ just after the $k = 2$ mode pulse starts to decay. This implies that we can "predict" which shall be the next mode that produces the next pulse from the calculation of the intensity circulation. This was pointed out also for APS in modulated multimode lasers [12]. The phase portrait in the $[I_k, G_k]$ plane and state points of individual modes on the trajectory at a particular time are depicted in Fig. 2.28(c), where the moving direction of state points are indicated by arrows. In APS states, different modes find themselves on different positions along the same closed curve as shown in the figure.

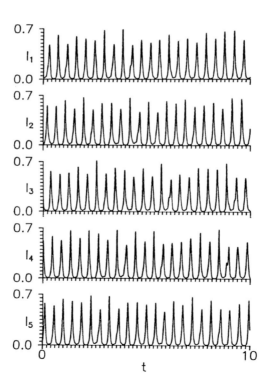

Figure 2.29: Local APS chaos. $\gamma = 0.0145$ and other parameters are the same as in Fig. 2.27.

2.3. CHAOTIC ITINERANCY IN ANTIPHASE

B. Transition from local APS chaos to global chaos via chaotic itinerancy

When small-signal gain γ is increased, the lasing intensity increases, the stable APS is destabilized and "local" chaotic APS as shown in Fig. 2.29 appears via quasiperiodicity. In this regime, each mode produces chaotic pulses, however, the ordering of the sequence does not change and the phase relationship among modes is approximately maintained during temporal evolutions. The upper bound of the fluctuation of pulsation periods around the long-term mean of the local chaotic APS periods is evaluated to be 20 percent.

As the small-signal gain is increased further, self-induced switching among the ruins of local chaotic APS attractors (chaotic itinerancy) occurs as shown in Fig. 2.30(a). In Fig. 2.30(a), switching $\{2, 1, 3, 4, 5\} \rightarrow$

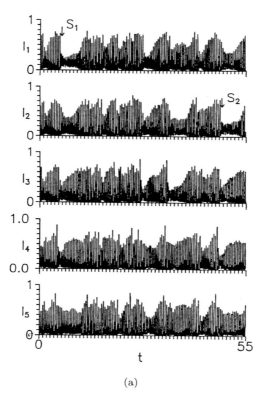

(a)

Figure 2.30: 4. Chaotic itinerancy among ruins of APS attractors. $\gamma = 0.015$ and other parameters are the same as in Fig. 2.27. (a) Temporal evolutions of modal intensities. (b) Enlargement of (a) near the switching point. (c) Corresponding intensity circulations from the $k = 1$ mode to all other modes and their summation.

$\{1,2,3,4,5\} \to \{2,1,3,4,5\}$ is clearly seen at points S_1 and S_2. The calculated Lyapunov spectrum (0.787, 0.396, 0.063, 0, −0.520, −0.789, −1.889, −2.513, −4.051, −8.138) contains three positive exponents. This implies that the motion is hyperchaotic (more than one positive Lyapunov exponent). Various switchings among different APS patterns, which are not restricted to the 'nearest neighbor pulses' like Fig. 2.30, are confirmed to occur.

To understand the switching process, we show modal intensities I_k and intensity circulations $I_{1,j}$ ($j = 2, 3, 4, 5$) near the switching point S_1 in Figs. 2.30(b) and 2.30(c). Before the switching point, all modes keep their APS phase relationship. When the system approaches the switching point, the intensities of all modes overlap to some extent at point F_1 as is seen in Fig. 2.30(b). At this point, simultaneous intensity flows from the $k = 2, 3, 4, 5$ modes to the $k = 1$ mode are created, as is identified from

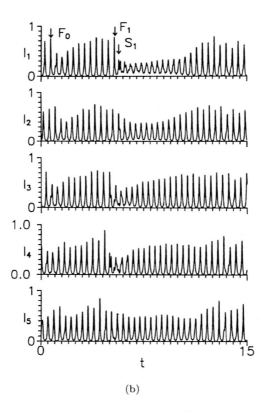

(b)

Figure 2.30: (continued).

Fig. 2.30(c). In short, all modes except the $k = 1$ mode transfer their intensities (i.e., power) cooperatively to the $k = 1$ mode. Such a cooperative power transfer, which creates an easy path for switching, is found to occur generally just before the switching point. The same calculations were carried out for $I_{2,j}(j \neq 2)$, $I_{3,j}(j \neq 3)$, $I_{4,j}(j \neq 4)$ and $I_{5,j}(j \neq 5)$ at point F_1, however, the cooperative intensity transfer was found to occur only for $I_{1,j}$. As a result, the $k = 1$ mode can receive power (energy) collectively from all other modes most efficiently at point F_1. This suggests that the $k = 1$ mode will be more strongly excited than usual. This parallels the result shown in Fig. 2.28(b). When the total intensity flow $\sum I_{1,j}(<0)$ exceeds a critical value, the cooperative intensity flow effect results in an excitation of the $k = 1$ mode at point S_1 after the sequence of $k = 1 \to 3 \to 4 \to 5$. At the same time the power flow from the $k = 1$ to the $k = 2$ mode peaks at switching point S_1. This implies that an exchange in the ordering of the $k = 2$ and $k = 1$ modes takes place and switching from $\{2, 1, 3, 4, 5\}$

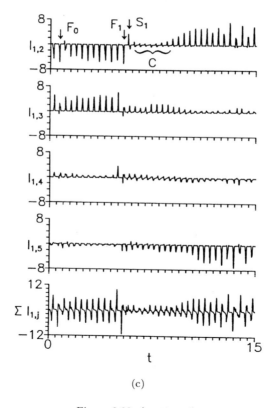

(c)

Figure 2.30: (continued).

to $\{1, 2, 3, 4, 5\}$ is established. At point F_0, on the other hand, the total intensity flow $I_{1,j}$ is smaller than at point F_1, the system fails to switch and remains in the $\{2, 1, 3, 4, 5\}$ sequence.

It should be noted here that in region C the $k = 1$ and $k = 2$ modes tend to approch the in-phase state transiently, in which $I_{1,2} \to 0$, while the other modes $\{3, 4, 5\}$ maintain the APS phase relationship. From the numerical simulations, such a state, in which two modes are in phase and the other modes exhibit APS, is found to form an unstable orbit existing at the basin boundary between APS's with different orderings of the sequence $\{I_k\}$.

To indicate the switching process more clearly, stroboscopic phase space trajectories of different modes on the $[I_k, G_k]$ plane during nearly one pulsation period and state points of different modes on the trajectories at a particular time before the switching (a) and after the switching (c) are shown in Fig. 2.31. The snap shots of state points of different modes are also shown. The switching between the mode 1 and 2 are easily seen. The phase space trajectories near the point F_1 and state points of different modes at the point F_1 are shown in Fig. 2.31(b). It should be noted that at the point F_1, the mode 1 on the attractor exhibits a high intensity value, while other modes show low intensities (see the snap shot at F_1 in Fig. 2.31(b)). This portrait paralles the cooperative unidirectional power

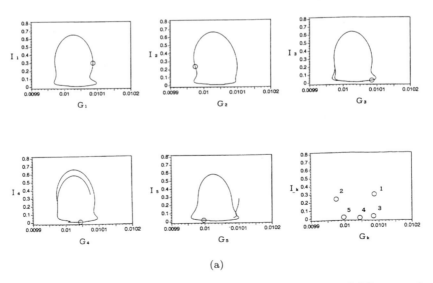

(a)

Figure 2.31: Stroboscopic phase space trajectories and state points of different modes at a particular time in the regime of chaotic itinerancy. Snap shots are also shown. (a) Before the switching. (b) Near the point F_1. (c) After the switching.

2.3. CHAOTIC ITINERANCY IN ANTIPHASE

flow to the mode 1 from all other modes at the point F_1, which is shown in Fig. 2.30(c).

When the small-signal gain is increased further, switching tends to occur more frequently; finally fully developed (global) chaos appears, as

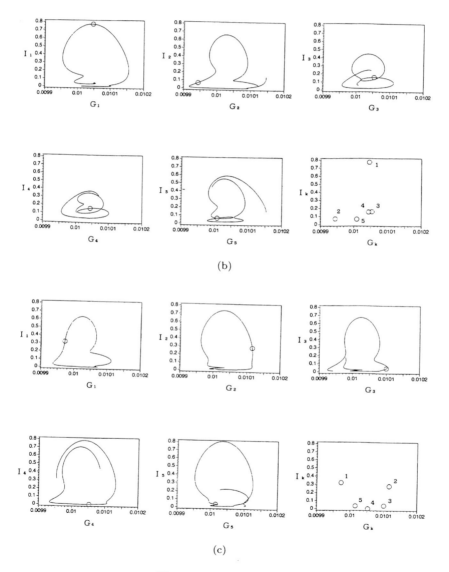

Figure 2.31: (continued).

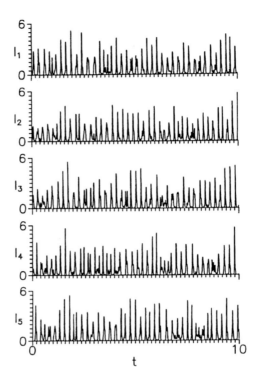

Figure 2.32: Global chaos at $\gamma = 0.03$. Other parameters are the same as in Fig. 2.27.

shown in Fig. 2.32, in which APS phase relationship is destroyed in addition to chaotic intensity evolutions.

The above scenario leading to global chaos via chaotic itinerancy that results from the rule of cooperative intensity flow occurs generally in the two polarization case as well.

A chaotic itinerancy among the ruins of local chaotic antiphase state attractors, resulting from a cooperative power flow, is now identified by intensity circulation analysis. The cooperative switching process resulting from the overlapping of modal intensities arises from the chaotic evolutional rule of the system itself, without external noise. In short, a quite ingenious mechanism which makes it much easier to change the ordering is self-generated through the chaotic dynamics. Such an itinerancy is not be brought about when random noise is applied to the system like Maxwell-Bloch turbulence described in **2.1**. Indeed, it is confirmed on the basis of numerical simulations that switchings never occur as a result of random noise applied to the gains of all modes in stable APS regimes. When the noise amplitude is increased, APS attractors are completely destroyed suddenly, leading to global chaos. This result implies that an internal force generated through a chaotic evolution of the system itself is more effective than an externally applied random force in searching for easy switching paths through which the itinerancy over the attractor ruins is realized. This is a common nature of the self-induced chaotic itinerancy phenomenon.

2.4 Clustering, Grouping, Self-Induced Switching and Controlled Dynamic Pattern Generation in an Antiphase Intracavity Second-Harmonic Generation Laser [D]

Clustering, grouping of antiphase motions and self-induced chaotic switching among factorial number of coexisisting grouping states, leading to antiphase periodic states, are described in this section in the reference model of ISHG in the regime of large conversion coefficient into second harmonic. Dynamical characterization of grouping behavior and self-induced switching is carried out by global intensity circulation analysis. Nonreciprocal independence of initensity flow in grouping states and switching-path formation process among coexisting grouping states are presented. The controlled switching-path formations between different periodic grouping states by perturbation-pulse injections is shown.

2.4.1 Introduction

In this section, several dynamical behaviors associated with Q-switching-type APS's are described in the model of intracavity second-harmonic generation in multimode lasers in the regime of large frequency conversion into second-harmonic. They include clustering, grouping of antiphase oscillations, self-induced chaotic switching among coexisting grouping states and intermittent chaotic antiphase motion in high-dimensional phase space. Dynamical characterization of the grouping state is described in terms of global circulation analysis [D-2]. Nonreciprocal independence of intensity flow in grouping states and switching-path formation process among factorial number of coexisting grouping states is identified numerically. "Timing"-controlled perturbation-pulse-induced flexible successive dynamical pattern generations utilizing coexisting periodic grouping states is demonstrated by numerical experiments.

2.4.2 Q-switching antiphase periodic states in ISHG

A. Antiphase periodic state (APS)

In the case of relatively small frequency-doubling coefficients, APS's consisting of "spiking-type" pulses are born via a Hopf bifurcation as described in the previous section **2.3**. In this case, APS's are destabilized when a small signal gain (e.g., pump parameter) is increased and local APS chaos appears, leading to global chaos. In local chaos regimes, antiphase relationships among pulsed modes are almost maintained, while modal intensities fluctuate chaotically. When global chaos appears, both modal phases and intensities fluctuate chaotically. In the transition process from local to global APS chaos, chaotic itinerancy takes place and self-induced switching among ruins of coexisting local APS chaotic attractors, which is driven by an internally generated cooperative energy flow, occurs. The generic feature of chaotic itinerancy is that it never occurs by an externally applied random force as presented in **2.3**.

In the regime of large frequency-doubling coefficients, on the other hand, "Q-switching-type" APS's consisting of square-wave pulses have been reported to appear when a pump parameter is increased [17]. There coexist $M_{APS} = (N-1)!$ equivalent APS attractors in the phase space. For brevity, following studies are carried in the case that all the modes are oscillating in the same polarization. The following results are obtained by numerical integration of the fundamental ISHG equations (2.30)–(2.31).

The threshold small-signal gain for the onset of APS is depicted in Fig. 2.33 as a function of number of oscillating modes (e.g., system size) N. Adopted parameter values are $\alpha = 0.02$, $\beta = 0.292$ and $\epsilon = 0.04$ reported in [19]. In the following simulations, these parameter values are fixed. As the system size becomes large, the threshold small-signal gain γ_{th} increases in a $\gamma_{th} \propto e^{3N/4}$ fashion for $N \geq 6$ and APS region deceases as well. This is reasonable since input energy shared by lasing modes decreases and the basin of attraction shrinks as N increases as well. A typical example of

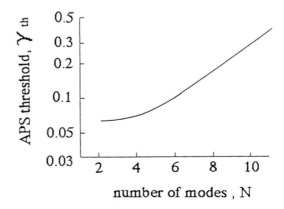

Figure 2.33: The threshold small-signal gain for APS, γ_{th}, as a function of system size N. Adopted parameter values are given in the text.

2.4. CLUSTERING, GROUPING, SELF-INDUCED SWITCHING

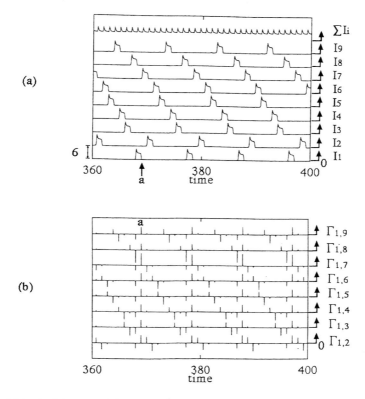

Figure 2.34: Antiphase periodic state. $\gamma = 0.3$. (a) Temporal evolution of intensities and their summation and (b) corresponding global intensity circulations $\Gamma_{1,j}$.

Q-switching-type APS's in the case of $N = 9$ is shown in Fig. 2.34(a), where $(N-1)! = 40{,}320$ equivalent APS's coexist in a high-dimensional phase space. It should be noted that the total output shows rather smooth amplitude oscillations superimopsed on a constant dc component (plateau).

B. Global intensity circulation

In Fig. 2.34(b), the global intensity circulations relative to the mode 1 is displayed. Here, the global intensity circulation is defined as [D-2]:

$$\Gamma_{ij} = \dot{I}_j \prod_{k \neq j} I_k - \dot{I}_i \prod_{k \neq i} I_k. \tag{2.35}$$

Here, the overdot denotes the time derivative. The product $\dot{I}_j \prod_{k \neq j} I_k$ represents the total intensity transfer from all modes to mode j. Γ_{ij} includes the contributions of *all* lasing modes instead of only modes i and j. Hence it describes the global intensity circulation between modes i and j. At the pulse trailing edge $t = a$, the larger intensity (e.g., energy) flow occurs from the mode 1 to the mode 7 than to other modes. As a result, the mode 7 is excited in the next period by obtaining a predominant energy. The same scenario is observed for the energy flows at trailing edges of each mode and the sequence of APS is determined.

2.4.3 Dynamical states associated with APS

In the case of larger system size (e, g, $N \geq 8$), various dynamical behaviors, such as clustering, grouping of antiphase motions and self-induced switching among grouping states, appear in pump regimes below the APS threshold. The result for $N = 9$ is summarized in Table 2.2. In this table, CS, GS, SS denote periodic clustered state, periodic grouping state and self-induced chaotic switching among grouping states, respectively. In the following, results for $N = 9$ are mainly shown. All dynamical states leading to APS's are observed for $N \geq 8$ and only SS was observed for $N = 6$ and 7. For $N \leq 5$, such dynamical states as CS, GS and SS are not clearly seen. In pump regimes above the APS region, intermittent APS states appear.

Table 2.2: Dynamical states at different small-signal gains γ. $N = 9$.

small-signal gain γ	dynamical states
0.12	CS[2,7]
0.125	CS[4,5]
0.151	GS[7,2]
0.17	GS[8,1]
0.20	SS
0.30	APS

A. Clustered state (CS)

When a pump parameter is far below the critical pump for the onset of APS, the energy supply to the laser medium is not enough for all the lasing modes to produce APS pulses and clustered states are formed instead. Examples of temporal evolutions of clustered states are shown in Figs. 2.35 ($[m, n] = [2, 7]$) and 2.36 ($[4, 5]$). In clustered states, m lasing modes emit small-amplitude "inphase" square-wave pulses while other n modes emit larger amplitude antiphase Q-switched pulses such that the total intensity fluctuates periodically around a larger dc component. In other terms, the sum of inphase periodic modes is antiphased with Q-switched pulses as shown in Figs. 2.35 and 2.36. A number of "inphase" pulses decreases as a pump parameter is increased. A similar clustering behavior is often

2.4. CLUSTERING, GROUPING, SELF-INDUCED SWITCHING

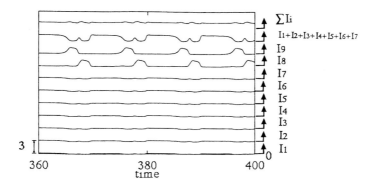

Figure 2.35: Clustered state [2, 7]. $\gamma = 0.12$.

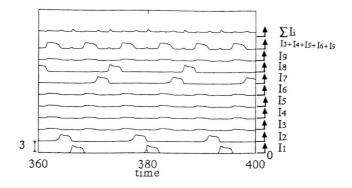

Figure 2.36: Clustered state [4, 5]. $\gamma = 0.125$.

observed in globally coupled laser systems, and it is referred to as AD2 state [18].

B. Grouping state (GS)

If a pump parameter is increased further, the energy supply to the laser increases accordingly. As a result, a number of Q-switched pulse modes increases. Then, another type of dynamical states of "grouping" takes place, in which a group of m modes smaller than N forms complete APS's, while other n modes exhibit higher-frequency periodic sustained relaxation oscillations (abbreviated as SRO) with smaller amplitudes. Therefore, the system is grouped into qualitatively different motions with different oscillation time scales. A number of SRO modes decreases as a pump parameter is increased. Examples of temporal evolutions of grouping states and cor-

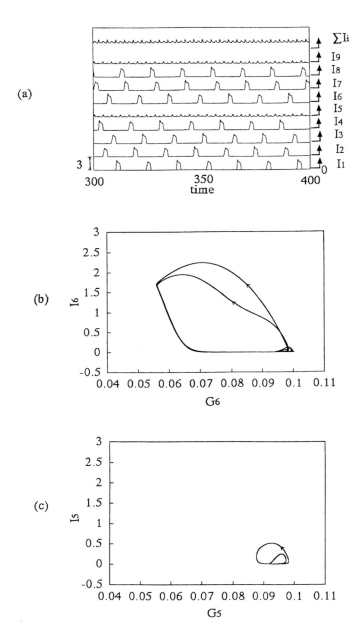

Figure 2.37: Grouping state [7, 2]. $\gamma = 0.151$. (a) Temporal evolution of intensities, (b) and (c) are phase space trajectories.

2.4. CLUSTERING, GROUPING, SELF-INDUCED SWITCHING

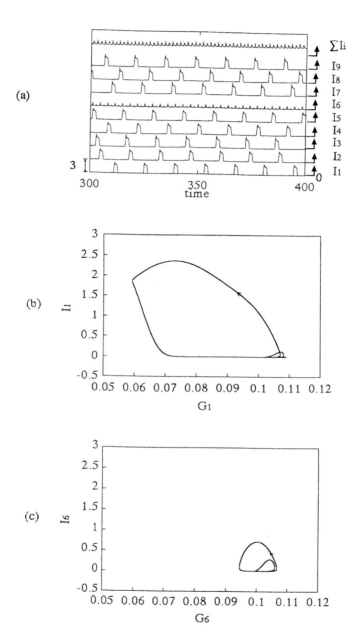

Figure 2.38: Grouping state [8, 1]. $\gamma = 0.17$. (a) Temporal evolution of intensities, (b) and (c) are phase space trajectories.

responding phase portraits are shown in Figs. 2.37 ($[m, n] = [7, 2]$) and 2.38 ($[8, 1]$), in which two or one mode(s) exhibit(s) SRO. In this case, there coexist $M_{GS} = N!(N - n - 1)!/n!(N - n)!$ equivalent grouping states in the phase space, where n is the number of SRO modes. Similarly to Fig. 2.35, the system is self-organized such that the total output exhibits rather smooth periodic fluctuations above a dc component, in which the total output oscillation waveform resembles that of SRO.

In this state, nonreciprocal independence of energy flow is found to be established. In short, a unidirectional global intensity flow occurs from the APS mode group to the SRO mode group. The global intensity circulation, which corresponds to Fig. 2.38(a), is shown in Fig. 2.39. Mode-to-mode global intensity circulations relative to the SRO mode fluctuate in both directions around zero, however, the sum of global intensity circulation indicates that a cooperative unidirectional energy flow from the APS mode group to the SRO mode is clearly seen to be established.

C. Self-induced switching among grouping patterns (SS)

When a pump parameter approaches the critical pump for the onset of APS, random switching among factorial number of coexisting grouping states appears. *It should be stressed that random switchings between different patterns take place at the same time of the occurrence of chaos.* Examples of switching patterns obtained for different initial conditions are shown in Fig. 2.40. A calculated Lyapunov spectrum possesses one positive, almost zero Lyapunov exponents (e.g., 0.24, 0.003) and 16 negative ones in the case of Fig. 2.40.

Depending on initial conditions, various switching patterns are generated after some transients, with different staying periods in the neighborhood of grouping state fixed points. In general, the chaotic SRO motion ap-

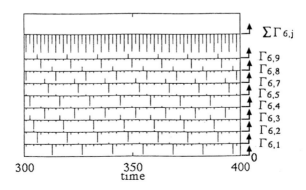

Figure 2.39: Unidirectional global intensity flow corresponding to the grouping state $[8, 1]$ shown in Fig. 2.38(a).

2.4. CLUSTERING, GROUPING, SELF-INDUCED SWITCHING 107

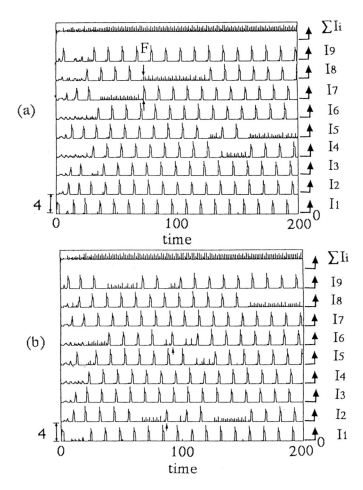

Figure 2.40: Self-induced switching states obtained for different initial conditions. $\gamma = 0.2$.

pears in a different mode at random in time like a "defect" wandering over oscillating modes. However, the total intensity still exhibits rather small-amplitude chaotic fluctuations superimoposed on a constant dc plateau, similarly to Figs. 2.37(a) and 2.38(a).

Example of self-induced switchings obtained for different system sizes are shown in Fig. 2.41.

There exists such a simple rule that switchings always occur quite 'locally' featuring an abrupt shift of a defect mode. At the point F in Fig. 2.40(a), for example, the chaotic SRO mode 7 produces a pulse just af-

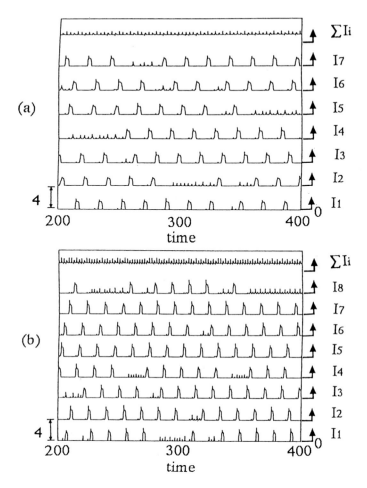

Figure 2.41: Self-induced switching obtained for different system sizes. (a) $N = 7$, $\gamma = 0.11$ (b) $N = 8$, $\gamma = 0.15$.

ter the preceding mode 6, as is indicated by ↑, while the mode 8, which was producing pulses after the preceding mode 6 before the switching, switches to a chaotic SRO state in return, as is indicated by ↓. As a result, the SRO mode shifts from 7 to 8 associated with switching. Here, other modes except for the switching pair modes 7 and 8 maintain their APS sequence for *at least more than one APS oscillation period* after the switching. The same switching pattern is established at other switching points in Fig. 2.40. In some cases, however, successive switchings occur *within one APS oscil-*

2.4. CLUSTERING, GROUPING, SELF-INDUCED SWITCHING

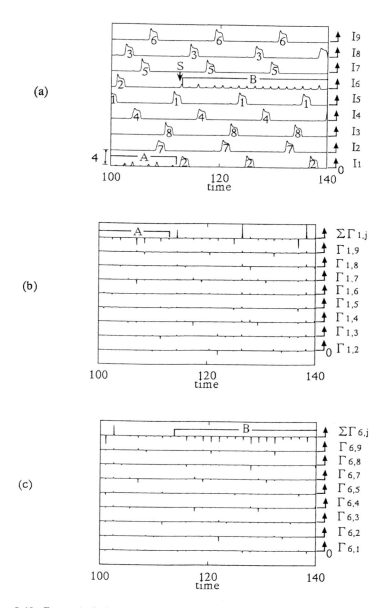

Figure 2.42: Dynamical characterization of self-induced switching. $\gamma = 0.2$. (a) Temporal evolution of intensities, where the switching sequence is indicated by numbers. (b) Global intensity circulation $\Gamma_{1,j}$ before the switching. (c) Global intensity circulation $\Gamma_{6,j}$ after the switching.

lation period obeying the above-mentioned rule as indicated by arrows in Fig. 2.40(b).

A cooperative unidirectional energy flow from the APS mode group to the SRO mode, which is shown in Fig. 2.39, is found to be occurring even in this chaotic domain. The result is shown in Fig. 2.42, together with corresponding temporal intensity evolution, where the sequences of APS before and after the switching point S is depicted by numbers. At the switching point, the global intensity flow in the system changes abruptly.

A clear theoretical explanation for the self-induced switching in this high-dimesional system has yet to be known, however, let us characterize dynamics occurring near the switching point to identify the switching mechanism numerically. For this purpose, global intensity circulations between the preceding chaotic APS mode 5 to the switching pair modes 1 and 6, and their difference are shown in Fig. 2.43. From this figure, it is found that a slightly larger energy flow to the mode 1 just before switching point S indicated by ↑ prefers the excitation of the mode 1 rather than the mode 6. Accordingly, the mode 1 produces an APS pulse, while the mode 6 switches to the chaotic SRO state in return. Additionally, energy flows from the preceding APS mode 5 to other modes except for the switching pair modes are found to be uniform and much smaller than to the mode 1. This parallels the APS pulse excitation mechanism in terms of energy flow shown in Fig. 2.34(b).

It should be stressed that local chaos, which is to be born from each periodic grouping state, is absent in this model. In modulated multimode lasers, on the contrary, local grouping chaos exists and a transient chaotic itinerancy among the ruins of local grouping chaos attractors terminated by a clustered state occurs as described in **2.2**. The absence of localized

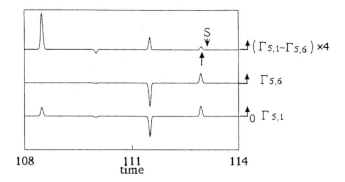

Figure 2.43: Global intensity circulations near the switching point in Fig. 2.42, involving the preceding mode 5 and switching pair modes 1 and 6.

2.4. CLUSTERING, GROUPING, SELF-INDUCED SWITCHING

chaos born from periodic grouping states may suggest that self-induced switching in the present system may be interpreted by a different context. This point will be discussed later.

D. Intermittent chaotic antiphase state

If a pump parameter is increased up to a critical point, all the lasing mode can share enough pump energy (e.g., population inversion) equivalently and realistic APS like Fig. 2.34(a) is established. As a pump parameter is increased above the APS region, APS pulses of individual modes exhibit intermittent destabilization as shown in Fig. 2.44. A calculated Lyapunov spectrum possesses 3 positive leading Lyapunov exponents indicating hyper-chaos. It is interesting to note that the present dynamical state is qualitatively different from other states in the sense that individual modes produce APS pulses with different shapes in time, whereas the total intensity exhibits intermittent spikes superimposed on a dc component.

Finally, it should be pointed out that the total output always exhibit rather smooth amplitude variations above a large dc component, while individual modes show a violent self-organized pulsations such that the laser produces more stable total output. Therefore, it is concluded that an equalization of intensities act as a strong constraint in the present system which governs the self-organized motion of individual modes for all kinds of dynamical states. In the case of small conversion into second-harmonic described in **2.3**, on the other hand, "sum rule" of $\sum_1^N I_i = $ constant has been reported [18].

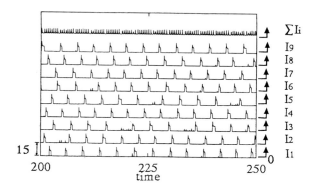

Figure 2.44: Intermittent chaotic antiphase state. $\gamma = 1$.

2.4.4 Perturbation-induced switching-path formation and controlled dynamic pattern generation

A. Switching-path formation by perturbation to SRO mode

There may exist a huge number of switching paths connecting multistable grouping states. The self-induced cooperative dominant energy flow to the chaotic SRO mode from the preceding APS pulse resulting in switching, which is identified in the previous section, suggests the possibility of controlled switching among grouping states by an external physical manipulation. In order to check the idea of "multiple" connections between coexisting grouping states and controlled switching, we apply an extremely weak light-pulse, which is 3 orders of magnitude smaller than the APS pulse height (i.e., $s_{i,k} \geq 0.002$, where pulsewidth $\Delta t = 1$), to a chaotic SRO mode as an 'external' perturbation instead of 'self-induced' dominant energy flow to a SRO mode and control the energy flows (e.g., switching-path formation) in regimes of *periodic* grouping states. From repeated numerical experiments, the following empirical rule is obtained.

Rule. Assume an APS mode sequence $\{\ldots, P, Q, \ldots\}$, where P and Q are arbitrary modes. If a perturbation pulse is applied to the SRO mode D at the time of leading edge of an APS pulse of the P-th mode, the SRO mode D emits an APS pulse after the preceding P-th mode instead of the Q-th mode and the Q-th mode switches to the SRO state in return, in which other modes maintain their APS sequence.

Results of light-injection-induced switching experiment are shown in Fig. 2.45, where a light-pulse with the same energy is applied to the periodic SRO mode 6 at T_1 (leading edge of the mode 7 pulse) and T_2 (leading edge of the mode 1 pulse), starting from the same initial condition. After a time delay on the order of the APS pulse period, the SRO mode produces APS pulses, while the mode 8 (see Fig. 2.45(a)) or the mode 5 (see Fig. 2.45(b)) switches to a SRO state in return, depending on the timing of the applied perturbation. As is stated in the rule, such switchings occur quite 'locally' and other modes except for switching pair modes (e.g., mode 6 and mode 8 in Fig. 2.45(a), mode 6 and mode 5 in Fig. 2.45(b), for example) maintain their APS sequence similarly to self-induced switchings shown in Fig. 2.40. This implies that $N - 1$ different switching-paths can be selectively created by changing the timing of the applied perturbation to the SRO mode within one APS oscillation, starting from the same grouping state. Note that such switchings does not occur if one applies "uniform" perturbation pulses to all the lasing modes simultaneously.

When a timing of perturbation is shifted from the leading edge and/or a pulse intensity is too strong, successive switchings indicated by arrows in Fig. 2.40(b) take place *within one APS oscillation period*, and finally an unexpected periodic GS pattern appears. In this case, externally-applied

2.4. CLUSTERING, GROUPING, SELF-INDUCED SWITCHING

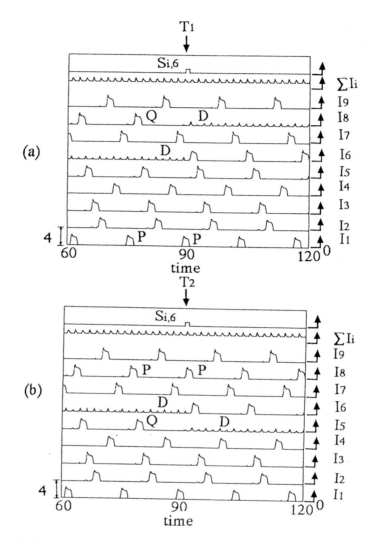

Figure 2.45: Perturbation-induced switching among periodic grouping states. $\gamma = 0.17$. The perturbation pulse intensity $s_{i,k} = 0.002$ and pulsewidth $\Delta t = 1$.

switching pulses act like random perturbations and one cannot predict a new pattern according to the above-mentioned rule.

B. *Controlled successive generation of factorial dynamic patterns*

If the initial periodic GS pattern is given, one can generate desired periodic GS patterns successively in a controlled manner by successive applications of switching pulses to SRO modes at different times according to the "program", which is based on the above-mentioned switching rule. In short, an extremely flexible successive excitations of a factorial number of dynamic patterns with weak perturbation signals is possible in the present system by a systematic physical manipulation based on the rule mentioned above. In the case of $N = 10$, for example, the maximum available patterns is $M_{GS} = 403{,}200$. An example of successive generations of periodic GS patterns is shown in Fig. 2.46. In this example, periodic GS pattern changes its APS sequence as $\{1,8,5,3,2,9,4,7\} \to \{1,8,6,3,2,9,4,7\} \to \{1,8,6,5,2,9,4,7\} \to \{1,8,6,5,3,9,4,7\}$ in accordance with the sequential shift of the "defect" mode $6 \to 5 \to 3 \to 2$. Random switchings are occurring when perturbation pulses with different intensities and phases are applied to individual modes at random. Note that in chaotic itinerancy systems, on the contrary, random switching among different attractors never occurs when a random external noise is applied as described in **2.1** and **2.3**.

If one tries to realize the "target" periodic GS pattern starting from a particular initial periodic GS pattern which has a large norm (distance) from the target pattern, there exist many switching routes reaching the target pattern via different intermediate GS patterns. Then, an interesting question arises as to find out the easilest switching route reaching the target pattern with the minimum number of perturbation signals.

The originality of this perturbation-induced switching is that the

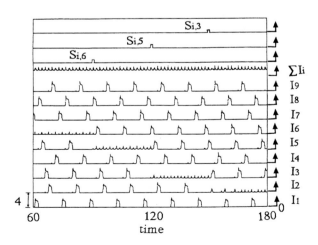

Figure 2.46: Successive generation of different periodic grouping states. $\gamma = 0.17$. $s_{i,k} = 0.002$, $\Delta t = 1$.

2.4. CLUSTERING, GROUPING, SELF-INDUCED SWITCHING

present autonomous antiphase dynamical system changes its dynamical pattern *by recognizing the "timing"* of an externally applied weak perturbation.

A clear understanding of peculiar self-induced switchings among coexisting grouping states has yet to be known theoretically, however, it is worth trying to give one of the plausible phenomenological explanations for the occurrence of self-induced switchings. The remarkable point of the present self-induced switching is the one-to-one correspondence between the occurrence of chaos and random switchings. This may suggest the possible existence of "heteroclinic" connections between coexisting grouping states. In other terms, an unstable manifold originating from one grouping state (e.g., saddle) is considered to cross a stable manifold of one of other grouping states. If such a crossing occurs transversally, it is known that the system switches between coexisting fixed points (e.g., grouping states) at random, at the same time, motion around grouping states exhibits very complex chaotic behavior.

Finally, it is worth mentioning that in the present reference model system, APS patterns survive stably for a large system size as compared

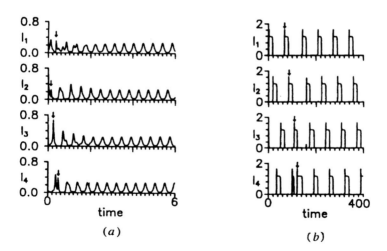

Figure 2.47: Assignment to desired APS pattern by injection-pulse seeding in ISHG systems. (a) Hopf-type APS state, $K = 5 \times 10^5$, $\beta = 0.6$, $\alpha = 0.01$, $\gamma = 0.01296$, $\epsilon = 5 \times 10^{-5}$ and $g = 0.52$. Four seed pulses are injected into the sequence $\{1, 3, 4, 2\}$ to produce the sequence $\{1, 4, 2, 3\}$. Seed pulse amplitudes are uniformly 5×10^{-5} and their duration is 0.012. The optimized position for the seed pulses is indicated by an arrow. (b) Q-switching-type APS state. $K = 500$, $\beta = 0.292$, $\gamma = 00.11$, $\epsilon = 0.05$ and $g = 0.5161$. Four seed pulses are injected into the sequence $\{1, 2, 4, 3\}$ at the times indicated by the arrows to produce the sequence $\{1, 2, 4, 3\}$. Seed pulse amplitudes are uniformly 2×10^{-5} and their duration is 1.

with other systems exhibiting APS. This implies that re-writable factorial dynamic pattern memory with an extremely large capacity of $C = \log(N-1)!/\log 2$ [bit] is feasible if a direct assignment to desired dynamical patterns by the injection-seeding method described in the previous section **2.2** is applied. Indeed, successful numerical demonstration of selective excitation of desired APS pattern by means of injection seeding has been reported in ISHG models described in **2.3** and **2.4** [D-2]. Examples are shown in Fig. 2.47.

The ISHG reference model in multimode lasers may be implemented in the form of globally coupled laser arrays including an intracavity period-doubling crystal [19].

Footnote

*1. *Chaotic Itinerancy (CI)*, that is self-induced switching among ruins of local attractors in high-dimensional chaotic systems, was reported for the first time in coupled nonlinear optical bistable systems [see Chapter **6**] by K. Otsuka, "Cooperative dynamics of nonlinear optical devices", at *Workshop on Applicability of Nonlinear Dynamics*, Advanced Telecommunications Research Institute International, Osaka, Japan 1988. Following this report, several CI papers in optics have been published: see for example, [A-1] K. Ikeda, K. Otsuka, and K. Matsumoto, *Prog. Theor. Phys. Suppl.* **99** (1989) 295; K. Otsuka, *Phys. Rev. Lett.* **65** (1990) 329; F. T. Arecchi, G. Giacomelli, P. L. Ramazza, and S. Residori, *Phys. Rev. Lett.* **65** (1990) 2531; T. Aida and P. Davis, *IEEE J. Quantum Electron.* **QE-30** (1994) 2986; K. Otsuka, T. Nakamura, J.-Y. Wang and J.-L. Chern, *Quantum and Semiclass. Opt.* **8** (1996) 1179.

CI has been identified in different physical and biological systems in recent years [for example, I. Tsuda, *World Futures* **32** (1991) 167 and K. Kaneko, *Physica* **D75** (1994) 55] and has been recognized to be a generic phenomenon in complex systems.

*2. *Spiking-Mode Oscillation* in deeply modulated class-B lasers was demonstrated for the first time by K. Otsuka in 1969 and was published in T. Kimura and K. Otsuka "Low-Frequency Resonance Phenomenon in Nd^{3+}:YAG Lasers", *Ooyou Butsuri*, Vol. 39 (1970) pp. 828-833 (in Japanese) [also see, T. Kimura and K. Otsuka, *IEEE J. Quantum Electron* QE-6 (1970) 764; K. Kubodera and K. Otsuka, *ibid.* **QE-17** (1981) 1139]. This was applied to semiconductor lasers for producing gigabit-rate picosecond optical pulse generations, which is now widely used as a pulse source for large-capacity optical transmissions, by S. Tarucha and K. Otsuka, *IEEE J. Quantum Electron.* **QE-17** (1981) 810.

REFERENCES

[Textbook]
[T] Paul Mandel, *Theoretical Problems in Cavity Nonlinear Optics*, Cambridge University Press, 1997. Chapters 7 and 10.

Keynote Papers
This Chapter is written on the basis of the following articles
[A]: Section 2.1
1. K. Ikeda, K. Otsuka, and K. Matsunoto, "Maxwell Bloch turbulence", *Prog. Theor. Phys.*, Vol. 99 (1989) pp. 295–324.
[B]: Section 2.2
1. K. Otsuka, "Winner-takes-all dynamics and antiphase states in modulated multimode lasers", *Phys. Rev. Lett.*, Vol. 67 (1991) pp. 1090–1093.
2. K. Otsuka, Y. Sato, and J.-J. Chern, "Grouping of antiphase oscillations in modulated multimode lasers", *Phys. Rev. A*, Vol. 54 (1996) pp. 4464–4472.
[C]: Section 2.3
1. K. Otsuka, T. Nakamura, J.-Y. Wang, and J.-L. Chern, "Chaotic itinerancy in antiphase intracavity second-harmonic generation", *Quantum and Semiclass. Opt.*, Vol. 8 (1996) pp. 1179–1188.
[D]: Section 2.4
1. K. Otsuka, Y. Sato, and J.-L. Chern, "Clustering, grouping, self-induced switching and controlled dynamic pattern generation in an antiphase intracavity second-harmonic generation laser," *Phys. Rev. E*, Vol. 56 (1997) pp. 4765–4772.
2. K. Otsuka, J.-Y. Wang, P. Mandel, and T. Erneux, "Dynamical characterization of globally coupled optical systems", *Quantum Semiclass. Opt.*, Vol. 7 (1995) pp. 461–466.

For information theoretic characterization of chaotic itinerancy, see J.-L. Chern and K. Otsuka, "Information theoretic consideration of the chaotic itinerancy in a globally coupled laser model", *Phys. Lett.*, Vol. A176 (1993) pp. 213–219.

References

1) H. Risken and K. Nummedal, J. Appl. Phys. **39** (1968) 4662.
2) R. Graham and H. Haken, Z. Phys. **213** (1968) 420.
3) P. Hadley and M. R. Beasley, Appl. Phys. Lett. **50** (1987) 621.
4) K. Yoshimoto, K. Yoshikawa, Y. Mori, and I. Hanazaki, Chemical Phys. Lett. **189** (1992) 18.
5) W. J. Freeman and C. A. Sarda, Brain Research Review **10** (1985) 147.
6) K. Kubodera and K. Otsuka, IEEE J. Quantum Electron. **QE-17** (1981) 1139.
7) W. Wiesenfeld and P. Hadley, Phys. Rev. Lett. **62** (1989) 1335.
8) C. L. Tang, H. Statz, and G. deMars, J. Appl. Phys. **34** (1963) 2289.
9) K. Otsuka and J.-L. Chern, Phys. Rev. A **45** (1992) pp. 8288.
10) P. Mandel, M. Georgiou, K. Otsuka, and D. Pieroux, Opt. Commun. **100** (1993) 341.
11) K. Otsuka, Gigabit Optical Pulse Generation in Integrated Lasers and Applications, *Picocecond Optoelectronic Devices*, Edited by Chi. H. Lee (Academic Press, Inc. 1984) and references therein.
12) K. Otsuka and Y. Aizawa, Phys. Rev. Lett. **72** (1994) 2701.
13) L. M. Pecora and T. Carrol, Phys. Rev. Lett. **64** (1990) 821; R. Roy and K. Thornburg, Jr., Phys. Rev. Lett. **72** (1994) 2009.

14) T. Baer, J. Opt. Soc. Am. B **3** (1986) 1175.
15) K. Wiesenfeld, C. Bracikowski, G. James, and R. Roy, Phys. Rev. Lett. **65** (1990) 1749.
16) J.-Y. Wang and P. Mandel, Phys. Rev. A **48** (1993) 671.
17) P. Mandel and J.-Y. Wang, Opt. Lett. **19** (1994) 533.
18) J.-Y. Wang, P. Mandel, and T. Erneux, Quantum and Semiclass. Opt. **7** (1995) 169.
19) J.-Y. Wang, P. Mandel, and K. Otsuka, Quantum and Semiclass. Opt. **8** (1996) 399.

Chapter 3

ANTIPHASE DYNAMICS IN MULTIMODE LASERS

This chapter presents antiphase dynamics and related phenomena which arise through the interaction of oscillating modes of class-B multimode lasers, in which polarization dynamics can be adiabatically eliminated. Among the vast number of class-B lasers, let me describe generic dynamics inherent in globally-coupled multimode lasers through the spatial hole-burning mechanism of population inversions discussed in the previous Chapter **2.2**, stressing experimental verifications in a microchip $LiNdP_4O_{12}$ (LNP) stoichiometric laser. Antiphase multimode laser dynamics which will be described in this chapter would provide prototypical examples of nonlinear dynamics in globally-coupled nonlinear oscillator systems.

3.1 Introduction

It has been known from the early dates of experimental observations soon after the invention of lasers such as a ruby laser that multimode lasers exhibit complicated behaviors. Complex multimode laser dynamics have recently been revisited as an intriguing subject for the study of nonlinear dynamics in optical systems.

The issue of intensity fluctuation in the output of lasers was initiated by the pioneering work of McCumber [1]. He predicted a noise peak in power spectra corresponding to relaxation oscillations in single-mode class-B lasers, in which polarization dynamics can be adiabatically eliminated, on the basis of linear response theory [1]. On the other hand, Tang, Statz, and deMars showed on the basis of numerical simulations for two-mode lasers with spatial hole-burning that the intensities of the individual modes show irregular spiking but the total output intensity could still exhibit a regular damped relaxation oscillation much as that of the single-mode laser [2].

In section **3.2**, we discuss a self-organized collective behavior in general N-mode free-running lasers and give an experimental evidence for that, featuring universal properties of multimode laser power spectra [A]. In section **3.3**, self-organized aspects featuring *antiphase dynamics* in unstable regimes of modulated multimode lasers are shown and the intermode statistical non-independence in chaotic multimode lasers following the vanishing gain circulation law is presented [B]. Multiple-parametric resonance effects in multimode lasers with multiple-frequency modulations are described in section **3.4** [C]. Finally, in section **3.5**, experimental observations of transverse effects on antiphase multimode laser dynamics and three-dimensional self-organizations in noise properties are described [D].

3.2 Self-Organized Relaxation Oscillations

In this section, let us examine self-organized collective behavior of free-running N-mode lasers and shows that an individual mode shows antiphase motion featuring N relaxation oscillation frequencies $f_1 > f_2 > \cdots > f_N$, while the total intensity exhibits a unique predominant relaxation oscillation, f_1, just like a single mode laser.

3.2.1 Multimode laser equations and linear stability analysis

A. *Tang-Statz-deMars (TSD) equations*

The scaled Tang-Statz-deMars equations for free-running multimode lasers with spatial hole burning [see **2.2**] are expressed by

$$dn_0/dt = w - n_0 - \sum_{k=1}^{N} \gamma_k(n_0 - n_k/2)s_k, \qquad (3.1)$$

$$dn_k/dt = \gamma_k n_0 s_k - n_k(1 + \sum_{i=1}^{N} \gamma_i s_i), \qquad (3.2)$$

$$ds_k/dt = K s_k[\gamma_k(n_0 - n_k/2) - 1], \quad k = 1, 2, 3, \ldots, N. \qquad (3.3)$$

Here, $w = P/P_{th}$ (P_{th}: first lasing mode pump threshold), n_0, n_k are the space average and the first Fourier components of population inversion normalized by the first mode threshold value $(N_0 - N_1/2)_{th}$, s_k is the photon density normalized by the steady-state S_1 value at $w = 2$, γ_k is the gain ratio to the first lasing mode and $K = \tau/\tau_p$ (τ: population lifetime, τ_p: photon lifetime). Time is scaled to the population lifetime, i.e., $t = T/\tau$.

The crucial role in the dynamics is played by the ratio of the cross-saturation parameter to the self-saturation parameter. In the present system, these coefficients are pump-dependent functions resulting from the fact

3.2. SELF-ORGANIZED RELAXATION OSCILLATIONS

that the spatial hole depth changes with pump intensity. They are always smaller than 1 and decrease with an increase of the pump, approaching $\gamma_k/2$ in the limit of $w \gg 1$.

B. Transient behavior

These equations are well recognized to express the transient behavior of multimode lasers governed by cross-saturation dynamics resulting from spatial hole-burning. We begin by showing a numerical result that clearly indicates self-organized transient relaxation oscillations. The result for five-mode lasers is shown in Fig. 3.1, assuming a gain (γ_k) distribution for oscillating modes. Here, the pump power is increased in a stepwise fashion from $w = 1.8$ to 2.1 at $t = 0.5$. The simulated waveforms show that the total output indicates a smooth relaxation oscillation with unique damping constant α_1 and angular frequency ω_1 just like a single mode laser, while individual modes feature other low frequency components. The ideal relaxation oscillation of the total intensity also implies that the oscillation components with frequencies lower than ω_1 seen in individual modes, which appear through cross-saturation dynamics, exhibit *antiphase* motion. This motion is such that these frequency components cancel each other out, yielding the ideal single relaxation oscillation at ω_1 for the total intensity. The low frequency antiphase motion exists even in the case of uniform gain, i.e $\gamma_k = 1$ independently of k. This implies that the antiphase motion arises through cross-saturation dynamics in lasers, since such motion never occurs for uniform gain in the absence of cross-saturation of population inversion.

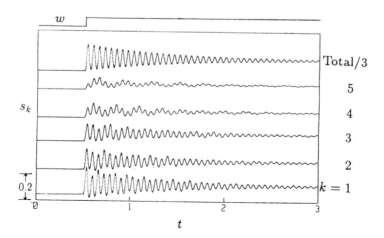

Figure 3.1: Simulated transient behavior of the intensities of a five-mode laser, where $K = 10^4$, $\gamma_5 = 0.94$, $\gamma_4 = 0.95$, $\gamma_3 = 0.97$, $\gamma_2 = 0.98$, and $\gamma_1 = 1$. The pump w is changed from 1.8 to 2.1 at $t = 0.5$.

C. Linear stability analysis [A-1]

Such a self-organized collective behavior has been proved analytically by a linear stability analysis for general N-mode lasers. First, let us examine some properties of stationary solutions. The non-trivial stationary solutions are most conveniently expressed in terms of n_0:

$$n_k = 2(n_0 - 1/\gamma_k), \tag{3.4}$$

$$s_k = (n_0 - 1/\gamma_k)/[\gamma_k R_1 - (N - 1/2)n_0\gamma_k]. \tag{3.5}$$

Here, $R_p = \sum_q (1/\gamma_p)^p$, $p = 1$ or 2, where the sum extends over N oscillating modes. The surprising property is that n_0 is the solution of the following equation, which is a quardratic for all N:

$$w = n_0 + (n_0 R_1 - R_2)/[R_1 - (N - 1/2)n_0]. \tag{3.6}$$

From these expressions, some analytic results are easily derived. For instance, n_0 varies from 0 to $R_1/(N - 1/2)$ as the pump w varies from 1 to ∞. The positivity of photon density s_k implies that a necessary condition for the k-th mode to oscillate is $\gamma_k > (N_k - 1/2)/R_1$ which, in return, can be shown to imply that γ_k must exceed $1/2$, where $N_k \leq N$. Thus the laser will operate on N modes only if $\gamma_N > (N - 1/2)/R_1$. Finally, the threshold for the N-th mode oscillation is given by $w_{th,N} = 1/\gamma_N + (R_1 - R_2\gamma_N)/(\gamma_N R_1 - N + 1/2)$.

In the case of single-mode oscillation, the simple linear stability analysis of the stationary solution yields the angular relaxation oscillation frequency:

$$\omega_1 = \sqrt{K(w-1)}, \tag{3.7}$$

To study the linear stability of multimode oscillations, two cases have to be considered separately, depending on whether the emerging mode is degenerate or not. Here, a mode will be q-degenerate if there are q indentical gain ratios corresponding to p different oscillation frequencies. The simplest case is $q = 2$ with two modes placed symmetrically with respect to the gain peak. We assume that the first $N - q$ modes oscillate stably and we consider the stability of the emerging p-degenerate modes. The algebra is quite tedius but straightforward. Hence, only the final results will be given. From the imaginary part of the characteristic equation, the angular relaxation oscillation frequency is given by

$$\omega_q = [(K/2)s_q(1 + qA/B)]^{1/2}, \tag{3.8}$$

where

$$A = \frac{2}{(\gamma_N)^{N-q}}(1 - 2\sum_{i=1}^{N-q}(1 - \gamma_N/\gamma_i)^2),$$

3.2. SELF-ORGANIZED RELAXATION OSCILLATIONS

$$B = \frac{1}{(\gamma_N)^{N-2-q}}[2(N-1)\sum_{i=1}^{N-q} n_i/\gamma_i - (N_0 + 2\sum_{i=1}^{N-q} 1/\gamma_i)(\sum_{i=1}^{N-q} n_i - n_0)].$$

However, if the threshold is q-degenerate, there will be in addition to the frequency (3.7) one, and only one, new frequency:

$$\omega_N = \sqrt{(K/2)s_N}, \qquad (3.9)$$

In the case of arbitrarily distributed γ_k, each time the threshold of oscillation of a new mode is reached, there are two new roots, one of which vanishes at the threshold. Slightly above threshold, the new roots become complex conjugate and a new relaxation oscillation given by Eq. (3.8) appears. As a result, $N-1$ different low frequency relaxation oscillations appear in addition to the McCumber frequency ω_1.

Next, let us examine total intensity dynamics. We first introduce new variables $\eta = t\sqrt{s_1^0 K}$, $s_k = s_k^0(1 + \bar{s}_k)$, $n_0 = n_0^0 + \bar{n}_0\sqrt{s_1^0/K}$, $n_k = n_k^0 + 2\bar{n}_k\sqrt{s_1^0/K}$, $\bar{s}_k = J_k + O(\epsilon)$, $\bar{n}_0^0 = N_0 + O(\epsilon)$, $\bar{n}_k = N_k + O(\epsilon)$, where the superscript "0" denotes the corresponding steady state solutions. In terms of these new functions, Eqs. (3.1)–(3.3) become

$$dJ_k/d\eta = (N_0 - N_k)(1 + J_k), \qquad (3.10)$$

$$dN_0/d\eta = -\sum_i^N J_i, \qquad (3.11)$$

$$dN_k/d\eta = (1 - n/2)J_k + (1 - n)\sum_{i=1, \neq k}^N J_i, \qquad (3.12)$$

where $n = n_0^0$ and $\gamma_k = 1 - O(\epsilon)$ has been assumed. A small amplitude analysis of these equations indicates that there is again one independent subset, $R_1 = \{N_0, \sum_{i=1}^N J_i, \sum_{k=1}^N N_i\}$ with eigenvalues $\lambda = 0, \pm i[2N - (N - 1/2)n]^{1/2} = 0, \pm\sqrt{K(w-1)}$. This suffices to prove that in the N-mode case, the rate equations predict a single frequency relaxation oscillation for the total intensity.

$$\omega_R = \sqrt{K(w-1)}. \qquad (3.13)$$

To summarize the results of linear stability analysis, the total output features damped relaxation oscillations at a unique frequency ω_R, while each mode may exhibit N relaxation oscillations, namely $\omega_R > \omega_2 > \omega_3 > \cdots > \omega_N$. This highest frequency ω_R is the McCumber frequency ω_1 of the single-mode laser given by Eq. (3.7).

3.2.2 Noise power spectra

To confirm self-organized relaxation oscillations and results on linear stability analysis in globally-coupled multimode lasers described in the previous section, the power spectrum properties of a free-running solid-state laser is investigated using a 1.32 μm cw LiNdP$_4$O$_{12}$ (LNP) microchip laser. The LNP crystal is $\ell = 1$ mm thick and both ends of the crystal are directly coated with dielectric mirrors (transmission at 1.32 μm of 0.1 and 1 percent). An argon laser ($\lambda = 514.5$ nm) serves as the pump. The 1.32 μm LNP laser radiation is linearly polarized along the pseudoorthorhombic b axis. Oscillation threshold is 130 mW and slope efficiency is 13 percent. Two-longitudinal-mode oscillation is obtained above $w_{th,2} = 1.09$ and three-longitudinal-mode oscillation is observed above $w_{th,3} = 1.7$. The light beam of the TEM$_{00}$ mode LNP laser is split into two beams by a beam splitter. One beam is passed through a spectrometer with a resolution of 0.1 nm, which is much smaller than the longitudinal mode spacing of the LNP laser of $\Delta\lambda = \lambda^2/2n\ell = 0.48$ nm ($n = 1.82$: refractive index).

The output from the spectrometer is detected with an InGaAs photodiode followed by a spectrum analyzer to observe the intensity fluctuation spectra of individual modes driven by a "white" noise. Another beam is used to measure the power spectrum of the total output.

Here, there is a one-to-one correspondence between the result of linear stability analysis and noise power spectra, which indicate a linear frequency response of the laser to a weak external perturbation [1]. Figure 3.2 shows the power spectra in the three-mode regimes measured by a spectrum analyzer for modal outputs (b), (c), (d) through a spectrometer together with a power spectrum of the total output (a) at $w = 2$ where the vertical scale for the total output is 5 times larger than those for modal outputs [A-2]. This figure clearly indicates that each mode shows three different frequency peaks f_1, f_2 and f_3 while the total output displays the highest frequency peak of relaxation oscillation at $f_1 = f_R$, in which low-frequency relaxation oscillation noise components are strongly suppressed. It is also seen that power spectral density of each mode P_k possesses the strongest noise component at f_k. To be more specific, $P_1 > P_2 > P_3$ at f_1, $P_2 > P_1 > P_3$ at f_2 and $P_3 > P_2 > P_1$ at f_3. Figure 3.3 shows noise power spectra calculated numerically by Eq. (3.1)–(3.3) including a "white" noise. The numerical results are found to reproduce experimental observations quite well. The power spectral density relation among modal output shown in Figs. 3.2 and 3.3 has also been derived analytically [A-3].

Universal relations which connect the power spectral density of each modal output and of the total output at the same frequency have been derived analytically [A-4] and they have been verified experimentally in a two-mode LNP laser [A-5]. Results are shown in *Supplement 1* in this

3.2. SELF-ORGANIZED RELAXATION OSCILLATIONS

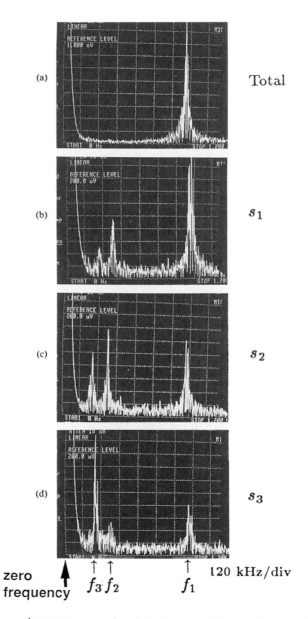

Figure 3.2: Measured power spectra of modal outputs and the total output of the three-mode LNP laser, where $w = 2$.

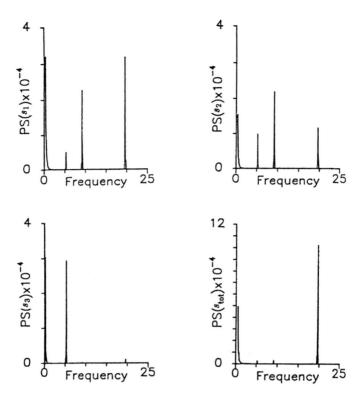

Figure 3.3: Numerical noise power spectra driven by a "white" noise for a three-mode laser. $w = 3$, $K = 10^4$.

chapter.

Figure 3.4 shows the relaxation oscillation frequencies measured as a function of the relative excess pump power $w - 1$, where K is estimated to be $120\mu s/310$ ps $= 3.9 \times 10^5$. When w is increased from zero, a single mode oscillation at wavelength 1.3217 μm appears at the threshold. Then, the second mode s_2 at 1.3222 μm goes into oscillation at $w_{th,2} = 1.09$ and the third lasing mode s_3 at 1.3211 μm appears above $w_{th,3} = 1.7$. It is found that $\Omega_\ell^2 \equiv \omega_\ell^2/K$ are well scaled by $w - w_{th,\ell}$, where $\ell = 1, 2, 3$. The solid line shows the theoretical value obtained for Ω_1 from the approximate equation (3.7), assuming $K = 3.9 \times 10^5$. Dashed lines are theoretical values obtained numerically from a small-signal frequency response of Eqs. (3.1)–(3.3), assuming $\gamma_2 = 0.97$ and $\gamma_3 = 0.9$. It is interesting to note that the lower frequencies are found to be well approximated by the following

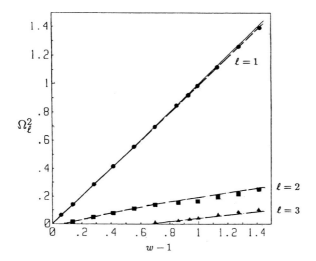

Figure 3.4: Measured relaxation frequencies as a function of the relative pump power of the LNP laser, assuming $K = 3.9 \times 10^5$. The solid line indicates the approximate Ω_1 value. The dashed lines are theoretical values numerically obtained by a small-signal pump modulation response of Eqs. (1)–(3), assuming $\gamma_2 = 0.97$ and $\gamma_3 = 0.9$.

relation [A-2]:

$$\Omega_\ell^2 = s_1 s_\ell \varphi(\gamma_\ell, n_0)/\Omega_1^2, \quad \ell = 2, 3, \ldots, N, \tag{3.14}$$

$$\varphi(\gamma_\ell, n_0) = -(\gamma_\ell - 1)^2 + \gamma_\ell n_0(1 + \gamma_\ell - 3\gamma_\ell n_0/4) = O(1). \tag{3.15}$$

From these investigations, a self-organized collective behavior in multimode lasers with spatial hole-burning is found to exist in general N-mode multimode lasers. The key is the antiphase cross-saturation dynamics, featuring low frequency components, among oscillating modes resulting from spatial hole-burning effect. Intrinsic low frequency noise peaks are absent in lasers without cross-saturation among modes such as semiconductor lasers.

3.3 Antiphase Dynamics in Modulated Multimode Lasers

In the previous section, antiphase dynamics in globally-coupled free-running multimode lasers was described in the linear regime, in which individual modes are self-organized such that the total output behaves just like single mode lasers. Then, the question arises: *Does such a self-organization character persist in nonlinear regimes?* To answer to this question, modulation dynamics are discussed in this section.

3.3.1 Modulation experiments

Modulation experiments are carried out by coupling the lasing beam from an argon-laser-pumped 1 mm flat LNP microchip solid-state laser, which is used in the experiment in the previous section, to a rotating paper sheet [3]. A schematic illustration of the experimental setup is shown in Fig. 3.5. It is well known that the photon statistics of the scattered-light field from a rough surface show a narrow-band Gaussian distribution whose center frequency is shifted owing to the Doppler effect [4]. Here, $\Delta f \ll f_D$ (Δf is the frequency linewidth of the scattered field, f_D is the Doppler-shift frequency) in this experiment. As a result, each lasing mode of the laser is found to be loss-modulated effectively at a Doppler shift frequency f_D resulting from interference between a lasing field and an extremely weak scattering field (self-mixing laser Doppler velocimetry scheme; SMLDV [5-7]).

To show the key dynamics, results for the simplest case of $N = 2$ and $N = 3$ are described here [B-1, B-2]. However, essentially the same collective behavior featuring antiphase dynamics is observed experimentally for larger number of modes.

Let us see the dynamics of the LNP laser modulated by the self-mixing laser Doppler velocimetry scheme mentioned above. The LNP laser is loss-modulated at the Doppler-shift frequency $f_D = 2v \cos \theta_s / \lambda$, where v is the angular velocity and θ_s is the angle between the laser axis and the velocity vector[3, 5–7]. When the intensity of injected scattered light is weak, clear resonances around $f_D = f_2$, $f_D = f_1/2$, $f_D = (2/3)f_1$, and $f_D = f_1$ are observed in the case of $N = 2$. At the same time, resonance occurs around $f_D = 2f_1$. Clustered states featuring antiphase motions are observed by

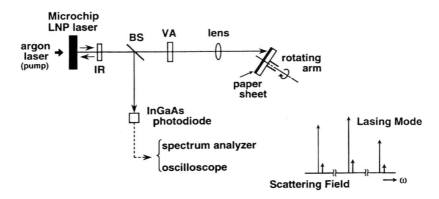

Figure 3.5: Schematic illustration of modulation experiment utilizing self-mixing laser Doppler velocimetry (SMLDV).

modulations around $f_D = f_2, f_3, \ldots, f_N$ for $N \geq 3$ mode regimes. The clustered states mean that oscillating modes are divided into several groups that exhibit different synchronized motions [See **2.2**]. Above f_1, the laser exhibited periodic oscillations at f_D over a wide range of velocities [5–7].

The modal output waveforms through a spectrometer and total output waveform at the f_2 and $f_1/2$ resonances for $N = 2$ are shown in Fig. 3.6(a) and (b) together with those for $f_D > f_1$ (Fig. 3.6(c)), where $w = 1.43$ and the power impinging on the paper sheet is attenuated such that a reflected scattering field amplitude $E_{k,s}$ is more than 5 orders of magnitude weaker than a lasing field amplitude inside the cavity E_k, i.e., $E_{k,s}/E_k < 10^{-5}$. In the case of the $f_D = f_2$ resonance, each output shows clear *antiphase* oscillations at the frequency f_2. For $f_D = f_1/2$, each mode exhibits period-2 sustained relaxation oscillations, in which the large peak in one mode corresponds to the small peak in the other, indicating antiphase dynamics. The same relationship is observed in the $f_D = (2/3)f_1$ periodic oscillation: two successive large peaks and one small peak in one mode are associated with two small peaks and one large peak in the other mode. Antiphase f_D

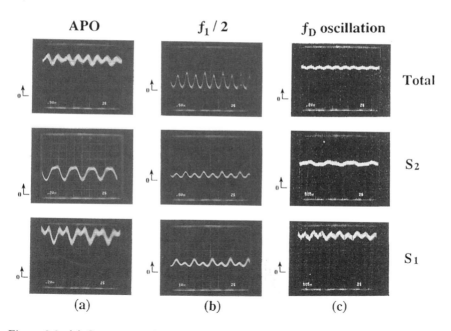

Figure 3.6: (a) Output waveforms indicating antiphase oscillations at $f_D = f_2$, where $w = 1.43$. 2 μs/div. s_1: first lasing mode intensity, s_2: second lasing mode intensity. (b) Output waveforms indicating antiphase period-2 relaxation oscillations at $f_D = f_1/2$, where $w = 1.43$. 2 μs/div. (c)Output waveforms indicating antiphase f_D (>f_1) oscillations, where $w = 2$. 2 μs/div.

oscillations are seen in Fig. 3.6(c).

For $N = 3$, clustering takes place at $f_D = f_2$ in which two modes s_2, s_3 exhibit in-phase periodic oscillations while s_1 shows antiphase motion against them as shown in Fig. 3.7. Similar clustering $[s_1, (s_2, s_3)]$ is observed for $f_D = f_3$. All the modes show synchronized relaxation oscillations for $f_D = f_1$ independently of N.

As the reflected scattering light intensity is increased, period-doubling oscillations leading to chaotic relaxation oscillations are observed by the modulation at $f_D \simeq f_1$. To find the interplay between oscillating modes in chaotic regimes, the injected light intensity is increased and power spectra of chaotic relaxation oscillations averaged over a long period of time are measured [12]. Example for $N = 2$ and 3 are shown in Figs. 3.8 and 3.9. Individual modes feature strong fluctuation components around antiphase motion frequencies $f_i (i = 2, 3, \ldots, N)$ in addition to f_1 and higher frequency components, while these low frequency antiphase motion components cancel each other out for the total intensity independently of N. This strongly implies that *a self-organized collective behavior based on antiphase dynamics is generic in multimode lasers with cross-saturation of population inversion and exists even in chaotic regimes*. In other words, a chaotic attractor of each mode inherits the antiphase dynamic character of the corresponding stationary state from which it is born through bifurcations. Therefore, oscillating modes are not statistically independent in their fluctuations. A related intermode statistical non-independence has been demonstrated in intracavity second-harmonic generation experiments on the basis of probability distributions of the output intensity of a chaotic

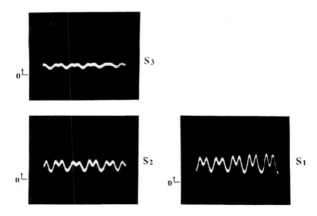

Figure 3.7: Output waveform indicating clustered state for $N = 3$ at $f_D = f_2$. $w = 2$. $2\mu s/$div.

3.3. ANTIPHASE DYNAMICS IN MODULATED MULTIMODE LASERS

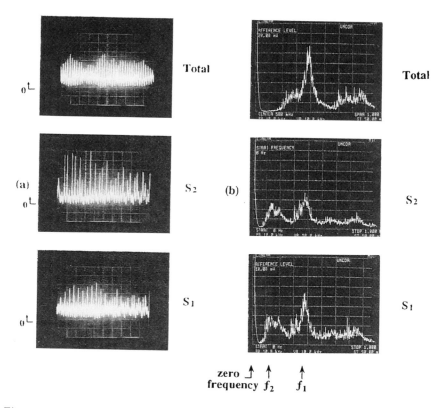

Figure 3.8: Output waveforms (a) and corresponding power spectra (b) indicating antiphase chaotic oscillations in two-mode regimes. $w = 1.4$. (a) 10 μs/div. (b) 100 kHz/div. The modulation frequency f_D is set near f_1. The power spectra are obtained by averaging "accumulated" power densities over 20–30 seconds. The measurements are not made simultaneously for two modes.

multimode laser [8].

3.3.2 Numerical analysis [B-1]

The observed phenomena including antiphase periodic and chaotic relaxation oscillations are well reproduced by numerical simulations of the model equations of multimode lasers with injected signals, in which individual modes are modulated at the Doppler-shift frequency and this 'local interaction' spreads over the entire system via global coupling among modes through cross-saturation of population inversion. Numerical simulations are carried out by adding a modulation term to Eq. (3.3) such

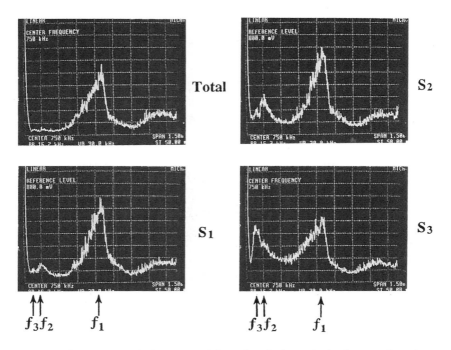

Figure 3.9: Power spectra indicating antiphase chaotic dynamics in three-mode regimes, where f_D is set near f_1. $w = 1.8$. 150 kHz/div. These power spectra are obtained by averaging "accumulated" power densities over 20–30 seconds.

that
$$ds_k/dt = K[\gamma_k(n_0 - n_k/2) - 1] + s_k Km \cos(\Omega_D t), \quad (3.16)$$

assuming $m = 2E_{k,s}/\kappa E_k$, κ is the coupling of field amplitude of the output mirror and $\Omega_D = 2\pi\tau f_D$ is the normalized Doppler-shift frequency, i.e., modulation frequency. Results based on Eqs. (3.1), (3.2) and (3.16) are shown in Fig. 3.10, assuming lifetime ratio $K = \tau/\tau_p = 3.9 \times 10^5$, and gain ratio of the second lasing mode to the first one $\gamma_2 = 0.97$. (In the case of a microchip laser with a short photon lifetime τ_p, K becomes extremely large and highly-sensitive modulation is found to be established from Eq. (3.16) even if the scattering intensity is extremely small [See *Supplement 2*]).

The resonant excitation of antiphase oscillations at $f_D = f_2$ and period-2 relaxation oscillation that are observed experimentally (see Fig. 3.6), are reproduced clearly (see Fig. 3.10(a) and (b)). The clustering behavior for $N \geq 3$ and antiphase dynamics in chaotic regimes are also reproduced by numerical simulations and power spectrum analyses of time series. An example of numerical results for clustered states is shown in Fig. 3.11.

3.3. ANTIPHASE DYNAMICS IN MODULATED MULTIMODE LASERS

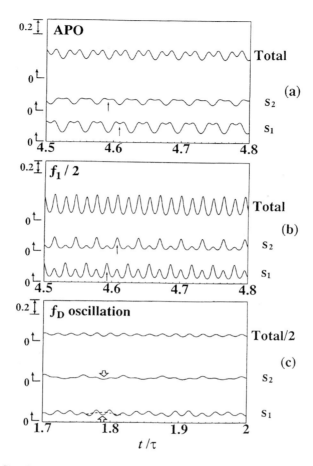

Figure 3.10: Simulated output waveforms for different f_D. $K = 3.9 \times 10^5$ and $\gamma_2 = 0.97$. The output intensities are scaled by steady-state S_1 value at $w = 2$. (a) $f_D = f_2$. $2\pi f_2 \tau = 183$, $w = 1.43$, and $\eta \equiv E_{i,s}/E_i = 10^{-5}$. Here, E_i and $E_{i,s}$ are lasing field and corresponding scattering field amplitude for i-th mode. (b) $f_D = f_1/2$. $2\pi f_1 \tau = 402$, $w = 1.43$, and $\eta = 5 \times 10^{-6}$. (c) $f_D > f_1$. $2\pi f_D \tau = 754$. $w = 1.64$, and $\eta = 2.5 \times 10^{-5}$. Antiphase dynamics are depicted by arrows.

3.3.3 Vanishing gain circulation [B-3]

To extract the underlying physics behind the self-organized collective behavior in globally-coupled multimode lasers which features antiphase dynamics, dynamical characterization is demonstrated in terms of the gain circulation among oscillating modes expressed by Eq. (2.28) in Chapter **2**. Numerical simulations indicate that the self-organization is established such

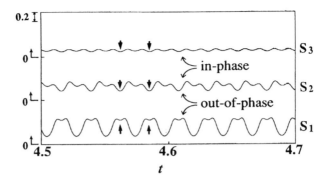

Figure 3.11: Simulated waveforms indicating $[s_1, (s_2, s_3)]$ clustered state in three-mode regime. $f_D = f_2$. $w = 2$. $2\pi f_2 \tau = 272$, $\eta = 2.5 \times 10^{-5}$. Other parameters are the same as Fig. 3.10.

that the gain circulations in all the closed gain flow paths involving arbitrary multiple lasing modes $3 \leq n \leq N$, $\sum G_{i,j}$, become negligibly small as compared with direct mode-to-mode gain circulation $G_{i,j}$ in all domains of oscillation states, i.e., transient behavior, antiphase periodic states (Fig. 3.6), clustered states (Fig. 3.7) and chaotic states (Figs. 3.8–3.9).

Numerical examples are shown in Figs. 3.12 and 3.13 for transient dynamics and modulated states (e.g., clustered state and chaotic state), respectively. In this simulation, a pump modulation $w(t) = w_0 + \Delta w_m \cos(\Omega_m t)$ is employed, however, the following results are obtained for loss modulation as well. From numerical simulations over all possible paths and for a uniformly scanned parameter region with explicitly given range of all parameters, the upper bound of the fluctuation amplitude of gain circulation along closed gain flow path p, $G_p(t) = \sum G_{i,j}$, is evaluated to be 10^{-3} times that of the average gain transfer along path p, $G_{av} \equiv < \sum G_{i \to j} >$, i.e., $R_p \equiv G_p/G_{av} < 10^{-3}$ even in strongly modulated regimes like Fig. 3.13, where $G_{i \to j} = g_i \dot{g}_j$. The upper bound value is much smaller in transient regimes like Fig. 3.12.

These results imply that the nonreciprocal local (i.e., mode-to-mode) gain transfers are self-organized so as to ensure reciprocal average gain flow among modes approximately and the temporal evolutions of individual mode intensities, which feature antiphase dynamics, are determined accordingly. In other terms, the existence of the one-to-one correspondence between vanishing gain circulation rule $\sum G_{i,j} \ll G_{i,j}$ and antiphase dynamics suggests the *dynamical equipartition of popupation inversions* among oscillating modes in all domains of dynamical states.

3.4. PARAMETRIC RESONANCE

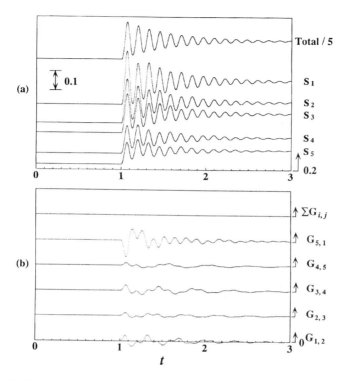

Figure 3.12: Simulated transient behavior of the intensities s_k of a five-mode laser: $\gamma_1 = 1, \gamma_2 = 0.98, \gamma_3 = 0.97, \gamma_4 = 0.95$ and $\gamma_5 = 0.94$. $\epsilon = 1.2 \times 10^{-7}$. The laser is initially in the stationary state for the pump power $w = 3$ and w is changed from 3 t0 3.5 at $t = 1$. (b) Temporal evolutions of gain circulation for (a). Gain circulations involving four modes ($k = 1, 2, 3, 5$) and the corresponding $\sum G_{i,j}$ are shown.

3.4 Parametric Resonance in a Modulated Microchip Multimode Laser

As described in previous sections in this chapter, globally-coupled N-mode multimode lasers with spatial hole-burning possess inherent N multiple relaxation oscillations. Then question arises: *What kind of behavior takes place when the laser is perturbed at these frequencies simultaneously in nonlinear regimes?* In this section, a nonlinear response of a laser-diode-pumped microchip $LiNdP_4O_{12}$ (LNP) multimode laser which is subjected to multiple-frequency modulations is presented [C-1, C-2, C-3]. Clustering and breathing motions featuring intermode parametric resonances are demonstrated when the LNP laser is modulated by rational frequencies chosen near-resonant to multiple relaxation oscillation frequencies inherent

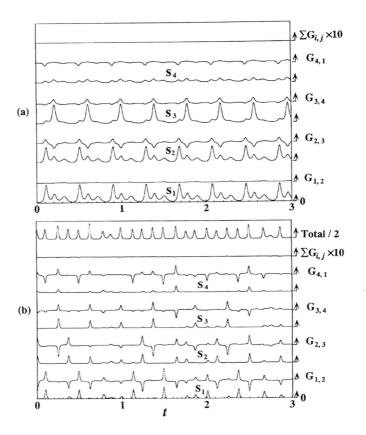

Figure 3.13: Temporal evolutions of the intensities and corresponding gain circulations for (a) the periodic clustered state and (b) the chaotic state. $N = 4$, $K = 1000$, $\epsilon = 1.2 \times 10^{-7}$, $\gamma_1 = 1$, $\gamma_2 = 0.97$, $\gamma_3 = 0.95$ and $\gamma_4 = 0.90$. (a) $w_0 = 3.5$, $\Delta w = 0.4$, $\Omega_m = 50$. (b) $w_0 = 3.5$, $\Delta w = 0.9$, $\Omega_m = 47$.

in multimode lasers. It is shown that different spatial pattern signals can be transformed into dynamic patterns. A simple correspondence between the modulation signal patterns and total output power spectrum patterns is established, implying that the present complex system is self-organized to exhibit the response expected in linear systems in nonlinear regimes.

3.4.1 Multichannel laser Doppler velocimetry modulation scheme

In this experiment, a highly sensitive modulation method by means of self-mixing laser Doppler velocimetry scheme descibed in **3.3.1** is used. The

3.4. PARAMETRIC RESONANCE

experimental setup is shown in Fig. 3.14. The experiment is carried out by a LNP solid-state laser. The LNP sample is cut off from an as-grown crystal (2cm-diameter × 4cm-long in size) and is polished into a 1-mm-thick platelet. Both ends of the crystal are directly coated with dielectric mirrors (transmission of 0.1 and 1 percent at 1.05 μm). A laser diode (Opto Power Corp. wavelength: 808 nm) serves as a pump. A collimated LD beam is passed through an anamorphic prism pairs and is focused onto the LNP crystal through an input mirror (transmission at λ = 808 nm is 95 percent) by a microscope objective lens. The absorption coefficient of the LNP crystal at λ = 808 nm is $\alpha_p \simeq 100$ cm^{-1} and the LD pump light is completely absorbed within the crystal length. The input-output characteristics is shown in Fig. 3.15. The lasing threshold is P_{th} = 150 mW and the slope efficiency is 20 percent. The linearly polarized 1048 nm LNP laser radiation exhibits the fundamental transverse mode oscillation for an entire pump region. As the pump power is increased, a number of oscillating modes is increased up to five modes.

The multiple-frequency modulation experiments are carried out by using a multichannel self-mixing laser-Doppler-velocimetry scheme (SMLDV) shown in Fig. 3.14. The lasing beam is split into two beams by a beam splitter. One beam is passed through a spectrograph with a resolution of 0.1 nm. The modal outputs from the spectrograph are detected by InGaAs photo-

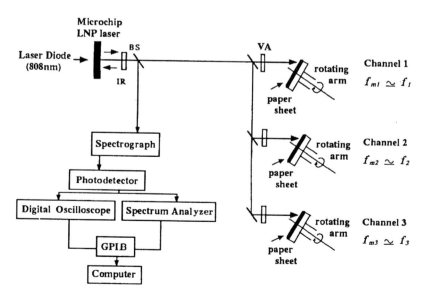

Figure 3.14: Experimental apparatus for multiple-frequency modulations. IR: infrared transmission filter, BS: beam splitter, VA: variable attenuator.

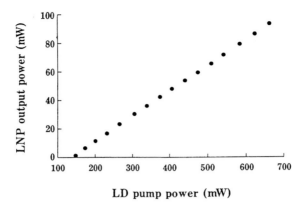

Figure 3.15: Input-output characteristics of a laser-diode-pumped LNP laser.

diodes, and modal output waveforms and corresponding power spectra are measured by the digital oscilloscope and the RF spectrum analyzer followed by a GPIB board and a personal computer. For the measurement of the total output, the spectrograph is removed. Another beam is split into multiple beams and they are impinged onto the different rotating paper sheets (multi-channel laser-Doppler-velocimetry feedback controller). Then, each lasing mode of the laser is loss-modulated effectively at Doppler-shift frequencies simultaneously resulting from the intereference of the lasing modal field and scattered fields from each channel as explained in **3.3.1**. The effective loss-modulation index is proportional to $K(E_{s,k}/E_i)$, where $K = \tau/\tau_p$ (τ: fluorescence lifetime; τ_p: photon lifetime), $E_{s,k}$ is the scattered field amplitude from the k-th channel and E_i is the lasing field amplitude of the i-th lasing mode. In the present microchip LNP laser, K has been evaluated to be 10^6 and highly-sensitive simultaneous loss-modulations at different frequencies is established although the scattered field amplitude is extremely small. Variable attenuators (VA) are inserted into each channel to change the scattered light intensity, which is fedback into the LNP laser, and to control the modulation index.

3.4.2 Periodic oscillations with multiple-parametric resonances

Experimental results for three-mode regimes obtained by using the three channel Doppler-feedback controller are shown. When the relative pump power is $w = P/P_{th} = 2.6$, three relaxation oscillation frequencies are measured to be $f_1 \simeq 1280$ kHz, $f_2 \simeq 540$ kHz, $f_3 \simeq 360$ kHz, respectively. Without modulations, modal output of the free-running laser, which is perturbed by a 'white noise', exhibited noise peaks at these frequencies, while

3.4. PARAMETRIC RESONANCE

the total output showed the noise peak only at f_1, where power spetrum peaks at f_2 and f_3 were strongly suppressed. Such a strong suppression of lower-frequency relaxation oscillations for the total output of multimode laser has been theoretically confirmed in linear regimes, in which multimode lasers are weakly modulated at resonant frequencies $f_1, f_2, \ldots, f_N$ simultaneously or are perturbed by a "white noise" as described in **3.2.1**. The linear regime is defined as the regime described by evolutional equations which are linearized around the steady sate.

Next, experimental results in nonlinear regimes with larger modulation indices are shown. The rotation speed of each rotator is set such that Doppler-shift frequencies of reinjected beams into the LNP laser become rational numbers which are near-resonant to multiple relaxation oscillation frequencies, f_1, f_2 and f_3. In the experiment, modulation frequencies $f_{m,k}$ are chosen to $f_{m,1} = 1280$ kHz, $f_{m,2} = 448$ kHz and $f_{m,3} = 320$ kHz. In short, the ratio of modulation frequencies is chosen to be 20/7/5. A variety of parametric mode interactions featuring clustering and breathing motions, which result from the rationality of the modulation frequencies, are observed according to different combinations of $[f_{m,1}, f_{m,2}, f_{m,3}]$. Hereafter, $[f_{m,1}, f_{m,2}, f_{m,3}] = [0\ 1\ 1]$ denotes that the $f_{m,1}$ beam is off (blocked) while the $f_{m,2}$ and $f_{m,3}$ beams are on (open), for example. In the case of $N = 3$, there are $2^3 = 8$ different periodic oscillations are obtained for parallel modulation signals including no-modulation case. In general, 2^N different parallel signals, e.g., spatial patterns, can be transformed into different dynamical patterns in N-mode lasers.

Examples of modal output waveforms and corresponding power spectra for the total intensity are shown in Fig. 3.16 where multiple-frequency modulations are applied to the LNP laser. In the case of [0 1 1] (Fig. 3.16(a)), the ouput shows a $[(s_1, s_2), s_3]$ clustered state with a 7/5 intermode parametric resonance, where a clustered state implies that the first lasing mode s_1 and the second lasing mode s_2 are in-phase while the phase of third lasing mode intensity s_3 is shifted from (s_1, s_2). In the case of [1 0 1] (Fig. 3.16(b)), the output shows a $[(s_1, s_2), s_3]$ clustered state with a 20/5 (= 4/1) parametric resonance. When the $f_{m,2}$ modulation is applied, long-term periodic breathing motions are observed. A breather mode oscillation featuring a 20/7 parametric resonance is obtained for [1 1 0] (Fig. 3.16(c)) and a breather mode oscillation with a 20/7/5 intermode parametric resonance is obtained for [1 1 1] (Fig. 3.16(d)). In the cases of single-frequency modulation [1 0 0], [0 1 0] and [0 0 1], in-phase sustained resonant relaxation oscillation, $[(s_1, s_2), s_3]$ and $[(s_1, (s_2, s_3)]$ clustered state are observed respectively, as described in **3.3.1**.

It is quite interesting to note that the power spectrum patterns (e.g., peak frequencies) for the total intensity and the modulation signal patterns show one-to-one correspondence as shown in Fig. 3.16. (Power spectra of

140   CHAPTER 3. ANTIPHASE DYNAMICS IN MULTIMODE LASERS

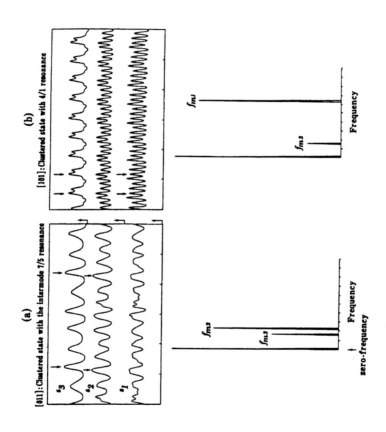

Figure 3.16: Experimental results indicating parametric resonances due to rational near-resonant multiple-frequency modulations. Modal output waveforms s_1, s_2, and s_3 were obtained by averaging over 100 acquisition time.

3.4. PARAMETRIC RESONANCE

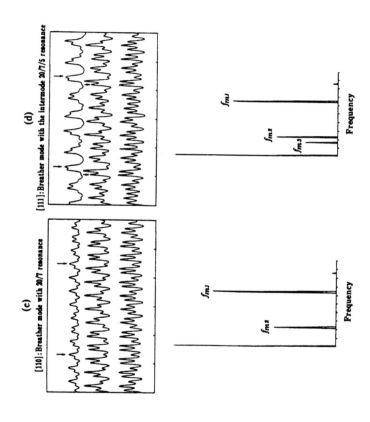

Figure 3.16: (continued).

modal outputs exhibited more complex properties.) In short, if an assignment of modulation frequencies $f_{m,k}$ to spatial channels is known, one can easily decode the spatial signal patterns by the power spectrum analysis of the total intensity. (In the case of Fig. 3.14, spatial assignment is: $f_{m,1} \to$ channel 1, $f_{m,2} \to$ channel 2, $f_{m,3} \to$ channel 3). Such a simple correspondence between the signal patterns and total output power spectrum patterns is established also for the single-frequency modulation cases. It should be pointed out that only the highest relaxation oscillation component at f_1 was observed for the total output in the case of linear regimes with the [1 1 1] modulation or with a "white" noise. (see Figs. 3.2 and 3.3, for example). The appearance of peaks at $f_{m,2}$ and $f_{m,3}$ in this case (see Fig. 3.16(d)) implies that the linear approximation is not sufficient to describe the experiment and that the modulation strongly affects the system.

3.4.3 Numerical verifications

The observed parametric resonances featuring clustering and breathing motions as well as the simple correspondence between modulation patterns and power spectrum patterns for the total output are well reproduced by numerical simulations of Tang-Statz-deMars equations for globally-coupled multimode lasers with spatial hole-burning including the following multiple-frequency loss-modulation term, i.e., (3.1), (3.2) and the following equation (3.17).

$$ds_k/dt = K[\gamma_k(n_0 - n_k/2) - 1] + s_k K \sum_{j=1}^{N} m_j \cos(\Omega_{m,j} t). \qquad (3.17)$$

Here, $m_j = 2E_{j,s}/\kappa E_j$ is the modulation amplitude and $\Omega_{m,j} \equiv \tau \omega_{m,j}$ is the normalized modulation frequency corresponding to Doppler shift frequency.

The observed phenomena are confirmed to take place generally independently of N. Examples of numerical results for the [011] and [111] modulations in the case of $N = 3$ are shown in Fig. 3.17(a) and 3.17(b), respectively. Here, adopted parameter values are $w = 2.6$, $K = 10^3$, $\gamma_2 = 0.97$, $\gamma_3 = 0.9$ and modulation frequencies are chosen to rational frequencies near relaxation oscillation frequencies, i.e., $2\pi\tau f_{m,1} \equiv \Omega_{m,1} = 40$, $\Omega_{m,2} = 14$ and $\Omega_{m,3} = 10$ with loss-modulation amplitudes $m_1 = 10^{-3}$, $m_2 = m_3 = 2 \times 10^{-2}$. Power spectra of modal outputs show complicated structures featuring several peaks, while power spectra of the total output exhibit the simple structure corresponding to the modulation signal patterns as demonstrated in the experiments.

3.4. PARAMETRIC RESONANCE

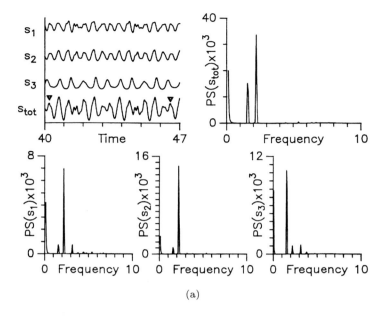

(a)

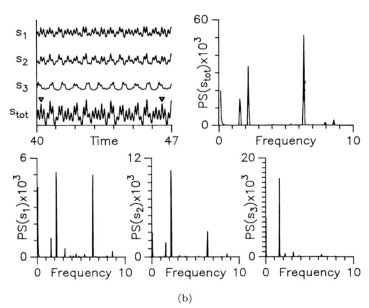

(b)

Figure 3.17: Numerical examples indicating parametric resonance effects for $N = 3$. (a) [011] modulation, (b) [111] modulation. The adopted parameter values are shown in the text. Time is measured in units of the population inversion lifetime τ.

From numerical simulations, the vanishing-gain-circulation rule, e.g., equipartition of population inversion among modes described in **3.3.3**, is found to hold for all the modulation patterns. In other words, this rule acts as the strong constraint which governs the dynamics even in nonlinear multiple-frequency modulation scheme.

The one-to-one correspondence between modulation patterns and power spectrum patterns for the total output implies the fact that the present globally-coupled *complex systems* is self-organized to exhibit a simple *linear response* which is expected in linear systems with multiple modulation signals, in nonlinear regimes such that the vanishing gain circulation law is established.

3.5 Transverse Effects on Antiphase Dynamics

The underlying physical processes inherent in self-organized antiphase dynamics and parametric resonance effects in globally-coupled *multi-longitudinal-mode* lasers are described and characterized in terms of gain circulation in previous sections in this chapter. In this section, experimental observations of *transverse effects* on antiphase dynamics in multi-longitudinal-mode lasers are discussed and "three-dimensional" self-organization is shown to result from the cross-saturation dynamics in the longitudinal and transverse directions [D-1, D-2].

The structure of the electromagnetic field in planes orthogonal to the direction of propagation in laser systems exhibits a rich variety of spatiotemporal phenomena. Important advances have recently been made in pattern formation and pattern dynamics in laser and nonlinear optical systems [9],[10], and this research subject is referred to as "transverse effects". In transverse laser dynamics, however, previous studies have been restricted to the dynamics of *multi-transverse modes* belonging to a single longitudinal mode.

The purpose of this section is to show experimental results on transverse effects on *multi-longitudinal-mode* antiphase laser dynamics. The observed generic feature is that a partial beam within the beam cross section features strong antiphase relaxation oscillation noise, resulting from the cross-saturation of population inversion in the longitudinal direction as described in **3.2**, and noise spectrum changes along the transverse direction such that the entire beam is completely free from antiphase relaxation oscillation noise in the case of simultaneously oscillating multiple transverse modes. This implies the existence of three-dimensional self-organization, in which the partial beam is noisy while the entire beam is quiet.

3.5.1 Transverse effect on noise power spectra

In this subsection, experimental transverse mode field effects on the antiphase laser dynamics are shown in a single TEM$_{00}$ as well as multiple transverse mode regimes. Experiments are carried out using a 1.32 μm cw LiNdP$_4$O$_{12}$ (LNP) laser. The LNP laser is the same as that in **3.2** and **3.3**.

The TEM$_{00}$ output beam is expanded and collimated to a parallel beam with a 5-mm spot size by a beam expander, and its intensity noise power spectra are measured at different positions within the beam using a detector with a diameter of 0.3 mm. It is found experimentally that self-organization resulting from cross-saturation dynamics in the longitudinal direction is established independently of the position within the beam cross section; the individual modes have different relaxation oscillation frequency peaks, $f_1 > f_2 > \cdots f_N$, while the total output indicates only the highest frequency peak of relaxation oscillation. In short, the noise power ratio, $R_\ell = I(f_i)/I(f_1)$ ($i = 2, 3, \ldots, N$) for each modal output, s_k ($k = 1, 2, \ldots, N$) does not depend on the position, while the total output is completely free from f_i components at all positions. Measured relaxation oscillation frequencies were found to coincide exactly with the theoretical values derived from the linear stability analysis of Eqs. (3.1)–(3.3) described in **3.2**. Space-dependent noise spectra of TEM$_{00}$ oscillation in a two longitudinal mode regime are shown in Fig. 3.18. This implies that Gaussian beam averaging is valid in the case of pure TEM$_{00}$ mode oscillation described in **3.2**–**3.4**.

When the Ar pump beam axis is tilted from the LNP lasing axis, the cylindrical symmetry is broken and higher-order Hermite-Gaussian oscillation modes appear. This is reasonable if we consider the overlap between the pump and lasing beam profiles [11]. In the case of pure higher-order transverse mode oscillations, e.g., TEM$_{10}$, TEM$_{20}$, ..., for instance, self-organization is also found to be established at all positions within the beam cross section. However, in the case of simultaneous oscillations of transverse modes, the situation changes drastically.

Simultaneous oscillations of transverse modes are easily obtained by controlling the tilt. Under the following experimental conditions described below, the lasing threshold for the TEM$_{00}$ mode is increased up to $P_{th} = 260$ mW and simultaneous oscillations of TEM$_{00}$ and TEM$_{10}$ are obtained above $w_c = P/P_{th} = 1.46$.

First, let us see position-dependent intensity noise power spectra for the total output in the regime where TEM$_{00}$ and TEM$_{10}$ modes belonging to a single longitudinal mode are oscillating simultaneously, i.e., [TEM$_{00}$, TEM$_{10}$] = [1, 1], where $w = 1.54$. The results are shown in Fig. 3.19, where (a) is the power spectrum for the entire beam, (b) is at the center

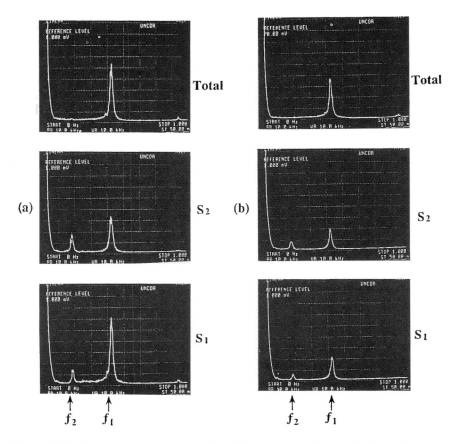

Figure 3.18: Measured power spectra of modal outputs and the total output of a free-running two-longitudinal-mode TEM$_{00}$ LNP laser (a) at the center of TEM$_{00}$ mode field profile, (b) at the foot of the TEM$_{00}$ mode field profile. 100 kHz/div.

of the TEM$_{00}$ field profile where the TEM$_{10}$ field is almost zero, (c) is near the peak of the TEM$_{10}$ field profile, and (d) is at the foot of the TEM$_{10}$ field profile where the TEM$_{00}$ field is almost zero. The oscillation mode spectrum, the far-field pattern and the intensity profile along the x-axis are shown at the left side of the figure. All the power spectra show a strong noise peak at the relaxation oscillation frequency f_1, but the low frequency noise peak at f_1^* appearing from the cross-saturation dynamics of population inversion in the transverse direction changes spatially, as shown in Figs. 3.19(b)–(d).

In this case, a strong position-dependent intensity overlap (i.e., cross saturation) exists between TEM$_{00}$ and TEM$_{10}$ modes along the transverse

3.5. TRANSVERSE EFFECTS ON ANTIPHASE DYNAMICS

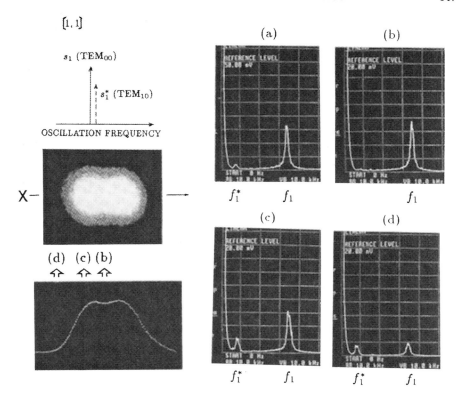

Figure 3.19: Position-dependent noise spectra of TEM_{00} + TEM_{10} mode LNP laser oscillation [1, 1] in free-running regimes, where $w = 1.54$. (a) entire beam (b) TEM_{00} mode profile center (c) near TEM_{10} mode profile peak (d) TEM_{10} mode profile foot. 100 kHz/div. A far-field pattern and its profile along X are shown on the left.

directions, while the intensity overlap between different longitudinal modes is independent on the position across the beam cross section for single-transverse-mode oscillations and the noise intensity ratio is constant. In Fig. 3.19, the noise intensity ratio, $R_t = I(f_1^*)/I(f_1)$, increases toward the foot of the TEM_{10} field profile. Figure 3.20 shows f_1 and f_1^* as a function of relative pump power $w - 1$. This figure suggests the following scaling relation, which is similar to the case of multi-longitudinal-mode oscillations described in **3.2** and **3.3**:

$$f_1^2 \propto (w - 1), \tag{3.18}$$

$$f_1^{*2} \propto (w - w_c). \tag{3.19}$$

It is apparent from Figs. 3.19(b)–(d) that the noise power spectrum for TEM_{10} possesses a stronger f_1^* component than TEM_{00}. Moreover, the

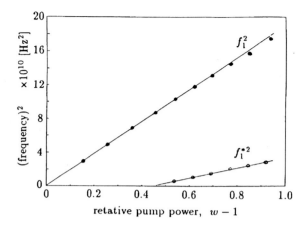

Figure 3.20: Measured relaxation oscillation frequenciès f_1 and f_1^* as a function of the relative pump power of the LNP laser.

f_1^* component is absent at the TEM$_{00}$ field center. Unlike the antiphase dynamics in the longitudinal direction, where low frequency relaxation oscillations cancel each other out in the total output, the low frequency relaxation oscillation component at f_1^* does not vanish, even for total output of the entire beam, as shown in Fig. 3.19(a). This may result from the difference in mode volumes due to the broken cylindrical symmetry and the strong position-dependent intensity overlap between the two transverse modes. In particular, the intensity overlap is almost zero around the peak of the TEM$_{00}$ mode and the foot of the TEM$_{10}$ mode.

3.5.2 Modulation dynamics

To confirm the different dynamic characters of two interacting transverse modes, modulation experiments are carried out using self-mixing laser-Doppler velocimetry described in previous sections. Figure 3.21 shows the output waveforms and the corresponding power spectra in the chaotic regime for $w = 1.62$ at different beam positions, in which [1, 1] scheme is established.

In the case of strong modulation at f_1, the output exhibits chaotic relaxation oscillations featuring broad noise around f_1 and higher harmonics, independently of the position in the beam. The fluctuation components around f_1^* are absent, even at the foot of the TEM$_{10}$ field, as shown in Figs. 3.21(a) and (a') (*transverse synchronization*). For the f_1^* modulation, on the other hand, the chaotic output in the TEM$_{10}$ foot features strong fluctuations around f_1^* in addition to f_1 and higher harmonics

3.5. TRANSVERSE EFFECTS ON ANTIPHASE DYNAMICS

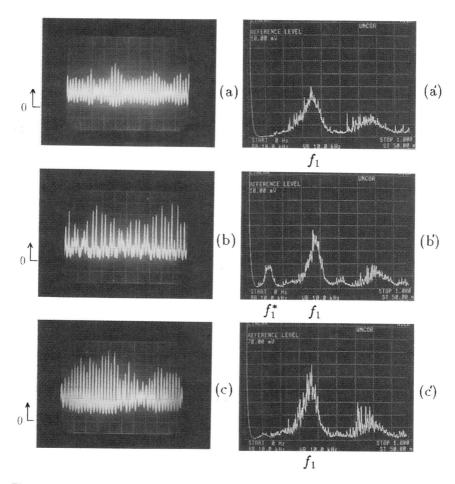

Figure 3.21: Position-dependent output waveforms and power spectra of a modulated $TEM_{00} + TEM_{10}$ LNP laser in the chaotic regime for the [1, 1] configuration at $w = 1.62$. (a), (a'): at the foot of the TEM_{10} field for f_1 modulation, (b), (b'): at the foot of the TEM_{10} field for f_1^* modulation, (c), (c'): at the center of the TEM_{00} field for the TEM_{00} field for f_1^* modulation. (a)–(c): 10 μs/div. (a')–(c'): 100 kHz/div.

(Figs. 3.21(b) and (b')), while the f_1^* component is absent in the TEM_{00} center (Figs. 3.21(c) and (c')) (*transverse clustering*). In this case, strong modulation at f_1^* induces chaotic relaxation oscillations of the TEM_{10} mode, and intensity fluctuation components around f_1 of the TEM_{10} excite chaotic relaxation oscillations of the TEM_{00} mode through transverse cross saturation.

Such transverse synchronization and clustering result from the fact that the f_1 component exists in both TEM_{00} and TEM_{10} modes, while the f_1^* component exists dominantly in the TEM_{10} mode, as shown in Fig. 3.19. These results strongly imply that the chaotic attractor of each transverse mode inherits the dynamic character of the stationary state (shown in Fig. 3.19) from which it is born through bifurcations.

In this experiment, the frequency difference between two modes is measured to be $f_b = 3.3$ MHz from the beat note and was much larger than the relaxation oscillation frequency f_1. As a result, dynamic instabilities resulting from phase-sensitive interaction among two transverse modes are not observed.

3.5.3 Three-dimensional self-organization of noise properties

Finally, we consider transverse effects on self-organization and antiphase dynamics in the regime where different longitudinal modes are oscillating by increasing the pump power. Figure 3.22 shows noise power spectra for the entire beam ((a) and (b)), at the center of the TEM_{00} mode field profile ((a') and (b')) and at the foot of the TEM_{10} mode profile for ((a") and ((b")) for [2,1] ($w = 1.77$) and [2,2] ($w = 1.85$) configurations.

It should be noted that the f_1^* component is absent at the center of the TEM_{00} field profile, similarly to Fig. 3.19(b), but the f_2 component resulting from the cross-saturation of population inversion in the longitudinal direction is present for the total output $s_1 + s_2$ even at the TEM_{00} mode field center, as shown in Figs. 3.22(a')–(b'). This is different from the case of single transverse mode operation, in which the power spectrum of the total output for the partial beam is completely free from antiphase relaxation oscillations, and is independent of the position within the beam cross section. This implies that self-organization due to the longitudinal cross-saturation is violated in the case of simultaneous transverse mode oscillations. The antiphase relaxation oscillations at f_2 and f_2^* due to $s_1 - s_2$ and $s_1^* - s_2^*$ interactions are clearly seen at the foot of the TEM_{10} mode field profile as shown in Figs. 3.22(a") and (b").

However, the antiphase relaxation oscillation components completely vanish for the entire beam, as shown in Figs. 3.22(a) and (b). This strongly suggests that the antiphase dynamic character changes qualitatively along the transverse direction, and that "three-dimensional" self-organization is established such that antiphase relaxation oscillation components resulting from cross-saturation dynamics in the longitudinal direction disappear completely when the entire beam is observed. In short, the partial beam is noisy, featuring antiphase relaxation oscillation noise, while the entire beam is quiet, being free from antiphase relaxation oscillation noise. This is of crucial importance to evaluating the noise properties of multimode

3.5. TRANSVERSE EFFECTS ON ANTIPHASE DYNAMICS

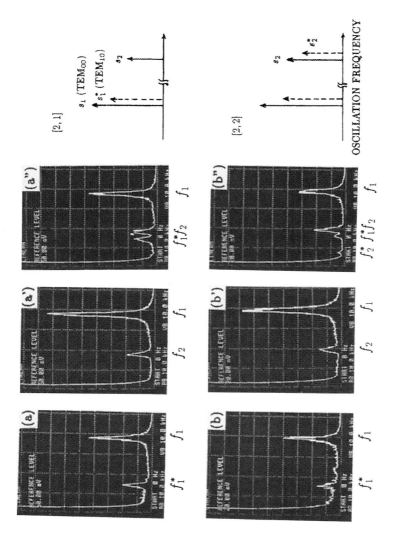

Figure 3.22: Position-dependent intensity noise spectra for entire beam ((a) and (b)), partial beam at the center of the TEM_{00} field ((a') and (b')), and at the foot of the TEM_{10} field ((a'') and (b'')) for a free-running LNP laser.(a)-(a''): [2,1] at $w = 1.77$, (b)-(b''): [2,2] regime at $w = 1.85$. 100 kHz/div.

lasers, and will provide a base for investigating dynamic pattern instabilities appearing from phase-sensitive interactions in $f_b < f_1$ regimes.

A theoretical study on the basis of longitudinal and transverse cross-saturation of population inversion and characterization of these effects in terms of three-dimensional gain circulation would be an interesting subject for the future study.

Supplement 1: Universal properties of multimode laser power spectra

According to the theory of multimode lasers expressed by the TSD equations [A-4], the relation between the peak of the power spectrum $P(s_n, \Omega_j)$ for mode s_n at frequency Ω_j and the peak of the power spectrum $P(\sum s, \Omega_j)$ for the total output $\sum s \equiv \sum_{n=1}^{N} s_n$ at the same frequency is given by

$$P(\sum s, \Omega_j) = \sum_{n=1}^{N} P(s_n, \Omega_j) \\ +2 \sum_{n=1}^{N} \sum_{m}^{n-1} \sqrt{P(s_n.\Omega_j)P(s_m,\Omega_j)} \cos(\varphi_{nmj}). \quad \text{(S.1)}$$

This result holds in the limit $1/K \to 0$. It involves the parameter $\varphi_{nmj} = \theta_{nj} - \theta_{mj}$, where θ_{nj} is the phase of n-th component of the eigenvector associated with the eigenvalue λ_j of the linearlized TSD equation around stationary states and Ω_j denotes its imaginary parts, i.e., relaxation oscillation frequencies.

In the simplest case of $N = 2$ (e.g, two mode oscillations), there are two relaxation oscillation frequencies, namely $\Omega_L < \Omega_R$ with $\gamma \equiv \gamma_2$ and $\gamma_1 = 1$, where $\Omega_R \equiv \omega_1$ and $\Omega_L \equiv \omega_2$ in **3.2**. It can be shown analytically that $\varphi_{12L} = \pi + O(1/K)$ and $\varphi_{12R} = O(1/K)$. Therefore the power spectra equalities become

$$P(\sum s, \Omega_R) = [\sqrt{P(s_1, \Omega_R)} + \sqrt{P(s_2, \Omega_R)}]^2, \quad \text{(S.2a)}$$

$$P(\sum s, \Omega_L) = [\sqrt{P(s_1, \Omega_L)} - \sqrt{P(s_2, \Omega_L)}]^2, \quad \text{(S.2b)}$$

from (S.1). These relations indicate antiphase dynamics. The universality of these relations, which do not require the knowledge of any laser parameter, stems from the fact that they relate peaks of different modal intensities at the *same frequency*. Hence they express a relation between different components of the same eigenvector and can therefore be independent of the preparation of the system.

These relations (S.2) for two-mode lasers have been quantitively verified both numerically and experimentally by using the 1.32 μm LNP laser

3.5. TRANSVERSE EFFECTS ON ANTIPHASE DYNAMICS

described in **3.2**, **3.3** [A-5]. An example of experimental noise power spectra is shown in Fig. S. Results obtained in the two-mode regime for different pump power (w) and gain (γ) are summarized in Table, where w_c is the threshold of the second lasing mode normalized by the first threshold. The excellent agreement is seen.

In a number of situations, the phase difference in (S.1) may be independent of the preparation of the system. In this case, the resulting

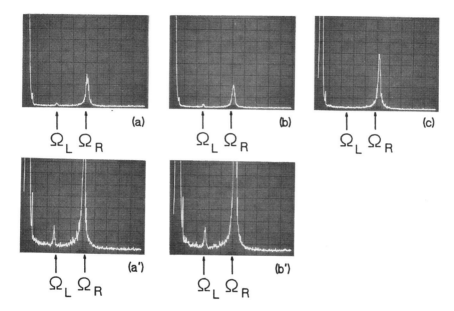

Figure S: Two-mode LNP laser power spectra in the free-running condition. (a) Power sperctrum for mode s_1. (a') Enlarged s_1 spectrum. (b) Power spectrum for mode s_2. (b') Enlarged s_2 spectrum. (c) Power spectrum for the total intensity $s_1 + s_2$, where the vertical scale is twice of (a) and (b). The horizontal scale is 100 kHz/div.

Table: Observed noise specrum normalized to $P(\sum s, \Omega_R)$ of free-running two-mode LNP laser. (a) $w = 1.6$, w_c (second mode threshold) $= 1.18$. (b) $w = 1.55$, $w_c = 1.26$. (c) $w = 3.27$, $w_c = 1.26$.

	Ω	$\sqrt{P(I_1,\Omega)}$	$\sqrt{P(I_2,\Omega)}$	$P(\Sigma I,\Omega)$ numerical	$P(\Sigma I,\Omega)$ calculated
(a)	Ω_L	0.22	0.26	< 0.01	< 0.01
	Ω_R	0.56	0.44	1	1.00
(b)	Ω_L	0.26	0.27	< 0.01	< 0.01
	Ω_R	0.54	0.43	1	0.94
(c)	Ω_L	0.16	0.16	< 0.01	< 0.01
	Ω_R	0.55	0.46	1	1.02

power spectrum relation becomes universal. This universality has been demonstrated analytically and experimentally in the case of $N = 3$, and distribution of the relative phase of modal intensities for arbitrary number of modes has been conjectured [A-3].

Supplement 2: Ultrahigh-sensitivity self-mixing laser Doppler sensing

Throughout this chapter, highly-sensitive self-mixing modulation of a microchip laser by a Doppler-shifted feedback light is featured. This technique is applicable to simultaneous measurement of different velocities of light scattering objects and vibration sensing [7]. The present SMLVD scheme with a laser-diode-pumped microchip laser possessing a large K-value (e.g., LNP) does not require sophisticated optics or electronics and makes easy remote sensing of multiple scattering objects simultaneously by using optical fibers. Indeed, a usable measurement is attained by reinjected light strength of less than 1 photon per Doppler-beat cycle, which is compatible to or less than that of the most sensitive photon-correlation LDV.

Keynote Papers

This Chapter is written on the basis of the following papers:
[A]: Section 3.2
1. P. Mandel, M. Georgiou, K. Otsuka, and D. Pieroux, "Transient and modulation dynamics of a multimode Fabry-Perot laser", *Opt. Commun.*, Vol. 100 (1993) 341–350.
2. K. Otsuka, M. Georgiou, and P. Mandel, "Intensity fluctuations in multimode lasers with spatial hole burning", *Jpn. J. Appl. Phys.*, Vol. 31 (1992) L1250-1252; K. Otsuka, P. Mandel, S. Bielawski, D. Derozier, and P. Glorieux, "Alternate time scale in multimode lasers", *Phys. Rev. A*, Vol. 46 (1992) 1692–1695.
3. P. Mandel, B. A. Nguyen, and K. Otsuka, "Universal dynamical properties of three-mode Fabry-Perot lasers", *Quantum and Semiclass. Opt.*, Vol. 9 (1997) 3655–380.
4. P. Mandel and J.-Y. Wang, "Universal properties of multimode laser power spectra", *Phys. Rev. Lett.*, Vol. 75 (1995) 1923–1926.
5. P. Mandel, K. Otsuka, J.-Y. Wang, and D. Pieroux, "Two-mode laser power spectra", *Phys. Rev. Lett.*, Vol. 76 (1996) 2694–2697.
[B]: Section 3.3
1. K. Otsuka, "Nonlinear dynamics in a microchip multimode laser", *Proc. SPIE on Chaos in Optics*, Vol. 2039 (1993) 182–197.
2. K. Otsuka, P. Mandel, M. Georgiou, and C. Etrich, "Antiphase dynamics in a modulated multimode laser", *Jpn. J. Appl. Phys.*, Vol. 32 (1993) L318–L321.
3. K. Otsuka and Y. Aizawa, "Gain circulation in multimode lasers", *Phys. Rev. Lett.*, Vol. 72 (1994) 2701–2704.
[C]: Section 3.4
1. K. Otsuka, D. Pieroux, and P. Mandel, "Modulation dynamics and spatiotemporal pattern generation in a microchip multimode laser", *Opt. Commun.*, Vol. 108 (1994) 265–272.
2. D. Pieroux, P. Mandel, and K. Otsuka, "Modulation dynamics in a modulated laser with feedback", *Opt. Commun.*, Vol. 108 (1994) 273–277.
3. K. Otsuka, D. Pieroux, J.-Y. Wang, and P. Mandel, "Parametric resonance in a modulated microchip multimode laser", *Opt. Lett.*, Vol. 22 (1997) 516–518.
[D]: Section 3.5

REFERENCES

1. K. Otsuka, "Transverse effects on antiphase laser dynamics", *Jpn. J. Appl. Phys.*, Vol. 10 (1993) L1414–L1417.
2. K. Otsuka, "Gain circulation and transverse effects in antiphase laser dynamics", *Chaos, Solitons and Fractals*, Vol. 4 (1994) 1547–1558.

References

1) D. E. McCumber: Phys. Rev. **141** (1966) 306.
2) C. L. Tang, H. Statz and G. deMars: J. Appl. Phys. **34** (1963) 2289.
3) K. Otsuka, IEEE J. Quantum Electron. *Vol QE-15* (1979) 655.
4) F. T. Arecchi, "Photocount distributions and field statistics", in *Quantum Optics*, R. J. Glauber ed. (Academic Press, New York and London, 1969), pp. 57-110; K. Otsuka, Appl. Phys. **18**, 415 (1979).
5) K. Otsuka, Jpn. J. Appl. Phys. **31** (1992) L1546.
6) K. Otsuka, Appl. Opt. **33** (1994) 1111.
7) K. Otsuka, OSA Proc. on Advanced Solid-State Lasers **15** (1993) 86.
8) C. Bracikowski and R. Roy, Phys. Rev. A **43** (1991) 6455.
9) L. A. Lugiato, Physics Reports **219** (1992) 293.
10) C. O. Weiss, Physics Reports **219** (1992) 311.
11) K. Kubodera, K. Otsuka and S. Miyazawa, Appl. Opt. **18** (1979) 884.

Chapter 4

MORE ON MULTIMODE LASER DYNAMICS

Following the previous chapter which focuses on well-recognized antiphase dynamics in multimode lasers, recent topical issues in multimode laser dynamics are discussed in this chapter and some puzzling problems are proposed for future studies. The power-law universality in intensity probability distribution of spiking laser oscillations and associate *nonstationary-to-stationary transition of chaos* are described in section **4.1** [A]. In section **4.2**, suppression of multimode chaotic spiking oscillations by means of *nearly resonant perturbation* on period-doubling bifurcation which appears as precursors to chaotic behavior [B]. In section **4.3**, chaotic burst generations in a multimode laser couped to an external cavity (e.g., single-mode optical fiber) are shown and they are interpreted in terms of a *frustration phenomenon* resulting from a double structure of "quantum" noise and "dynamical" instabilities [C]. Finally, breakup of multimode oscillations in a laser-diode pumped microchip solid-state laser by high-density pumping and an associate puzzling mode-partition instability are discussed in section **4.4**, featuring *quardratic-to-quartic transition of spatial hole-burning patterns* resulting from Auger recombination process among excited electrons [D].

4.1 Variation of Lyapunov Exponents on a Strange Attractor for Spiking Laser Oscillations

4.1.1 Introduction

Nonstationary deterministic chaos manifests itself in incomplete probability and is an intriguing aspect of complex dynamics. Good examples of this chaos include intermittent chaos and stagnation phenomena in Hamiltonian systems. In optics, chaotic spiking oscillations appear through deep pump and loss modulations near relaxation oscillation frequencies as well as through an incoherent delayed feedback in class-B lasers, where polarization dynamics can be adiabatically eliminated. (See Chapters **3.2**, **5.3**). The intensity probability distribution in such chaotic spiking oscillations $P(I)$ obeys the inverse power-law, i.e., $P(I) \propto I^{-1}$, in wide intensity regions. This inverse-power law universality is explained analytically in terms of a traditional giant-pulse model and is reproduced numerically. A typical example is shown in Fig. 4.1 for modulated chaotic spiking lasers (a) and for chaotic spiking lasers subjected to an incoherent delayed feedback (b) [see the next Chapter]. Figures 4.2(a) and (b) show an example of chaotic spiking oscillations and corresponding power spectra observed in a laser-diode-pumped $LiNdP_4O_{12}$ (LNP) laser with a deep pump modulation. The intensity probability distribution obtained from experimental data (Fig. 4.2(c)) is shown to obey the inverse-power law.

This inverse power-law property is associated with a universality class of phenomenon of *self-organized criticality* which has been investigated in modelling earthquakes, river networks, forest-fires, etc. This leads to the motivation for analyzing Lyapunov spectra in chaotic spiking lasers to provide understanding of temporal variations in local Lyapunov exponents on a strange attractor governing chaotic dynamics, in relation to the possible *nonstationary* characteristics in spiking lasers. Variation in local Lyapunov exponents is investigated by Abarbanel *et al.* [2]. These investigations showed on the basis of numerical simulations that variations about the mean of the Lyapunov exponents approach zero as the integral time $\tau \to \infty$ in the form of $\tau^{-\nu}$ ($\nu \simeq 0.5$–1.0) for several *stationary* chaotic systems, including the Henon map, Ikeda map and Lorenz model [2].

In this section, let us employ the standard deviation and Allan variance to characterize local variations of the growth rate of instabilities across the attractor for multimode spiking lasers. It is shown that local Lyapunov exponents along several expanding and contracting directions feature nonstationary variations in the time scale below the relaxation oscillation period τ_c, while variations approch zero as $\tau \to \infty$. This implies that qualitatively different variations exist in the local Lyapunov exponent, while ν is constant in entire τ regimes in standard chaotic systems [2]. The nonstationary variations of local Lyapunov exponents in short-term chaotic dynamics (i.e., short τ regimes) is confirmed by power spectrum analysis, where the power spectral density is given by $S(f) \propto f^{-\alpha}$ where $\alpha \geq 1$ in high-frequency regimes.

4.1.2 Local Lyapunov exponents and Allan variance

First, let us introduce the standard deviation and Allan variance, and test their usefulness for characterizing variation of local Lyapunov exponents on a strange attractor on the basis of Abarbanel *et al.*'s analysis [2].

4.1. VARIATION OF LYAPUNOV EXPONENTS

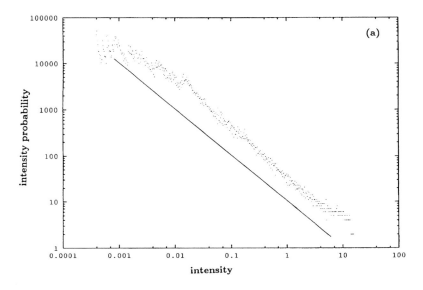

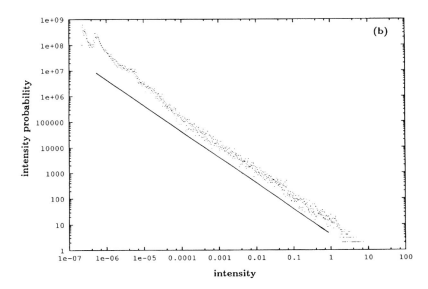

Figure 4.1: Chaotic spiking oscillations and intensity probability distribution. (Numerical results) (a) modulated laser, (b) laser subjected to incoherent delayed feedback.

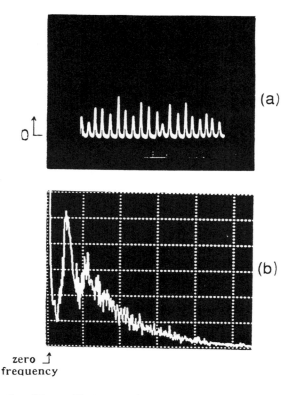

Figure 4.2: Chaotic spiking oscillation waveform of a pump modulated LNP laser (a) and power spectrum (b). The intensity probability distribution (c) obeys the inverse-power law.

Given a dynamical system in an n-dimensional phase space, the evolution of an infinitesimal n-sphere of initial conditions results in an n-ellipsoid. This is caused by the local deforming nature of the flow. The i-th one-dimensional local Lyapunov exponent is defined in terms of the length of the ellipsoidal principal axis $P_i(t)$:

$$\lambda_i(t) = \frac{1}{\Delta t} \log_2 \frac{P_i(t + \Delta t)}{P_i(t)}, \qquad (4.1)$$

where Δt is very small. The usual (global) Lyapunov exponent is thus

$$\Lambda_i = \lim_{T \to \infty} \frac{1}{T} \int_0^T \lambda_i(t) dt. \qquad (4.2)$$

4.1. VARIATION OF LYAPUNOV EXPONENTS

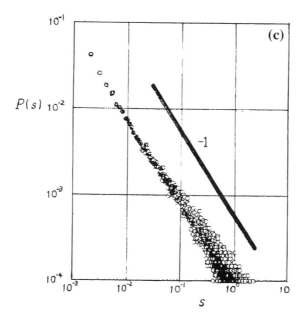

Figure 4.2: (continued).

The standard deviation μ_i and Allan variance σ_i^2 are defined as

$$\mu_i^2 = \lim_{N\to\infty} \frac{1}{N-1} \sum_{k=1}^{N-1} |\bar{\lambda}_{i,k} - m_i|^2, \qquad (4.3)$$

$$\sigma_i^2 = \lim_{N\to\infty} \frac{1}{N-1} \sum_{k=1}^{N-1} \frac{(\bar{\lambda}_{i,k+1} - \bar{\lambda}_{i,k})^2}{2}, \qquad (4.4)$$

where

$$\bar{\lambda}_{i,k} = \frac{1}{\tau} \int_{t_k}^{t_{k+1}} \lambda_i(t)dt, \quad k=1,2,\ldots,N, \quad t_{k+1} = t_k + \tau. \qquad (4.5)$$

Here, m_i is the mean value of local Lyapunov exponents λ_i. In calculations, $N \gg 1$ is assumed to be finite and the partial mean $m_i = \frac{1}{T}\int_0^T \lambda_i(t)dt$ is used, where $T \gg 1$ is finite.

Figure 4.3 shows the standard deviation and Allan variance as a function of the integral time τ for the stationary chaos, e.g., Henon map:

$$x(n+1) = 1 - ax(n)^2 + y(n), \qquad (4.6)$$

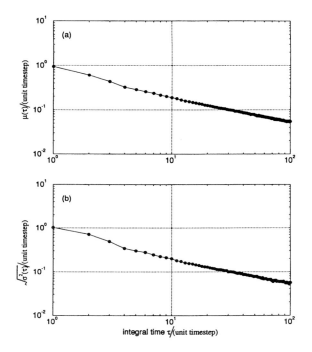

Figure 4.3: Standard deviation (a) and square root of Allan variance (b) for the larger local Lyapunov exponent variation in the Henon map.

$$y(n+1) = bx(n), \qquad (4.7)$$

assuming $a = 1.4$ and $b = 0.3$. The results are shown for the larger exponent. Data are collected from temporal evolutions of λ_i ranging from $n = 3000$ to 83000. The standard deviation, shown in Fig. 4.3(a), is found to decrease in the form of $\tau^{-0.5}$ as τ increases and corresponds well to Abarbanel et al.'s finding of $\simeq \tau^{-0.5}$ [2]. The scaling property of $\sigma^2 \propto \tau^{-1}$ in the Allan variance, shown in Fig. 4.3(b), also suggests a central limit theorem kind of variation, i.e., Gaussian Markov property.

4.1.3 Nonstationary chaos in spiking multimode lasers

A. Multimode spiking oscillation waveforms

Next, let us investigate the model of spiking multimode lasers shown in Fig. 4.2. For analysis, we use homogeneously broadened class-B lasers with spatial hole-burning [see **2.2** and **3.2**], such as the LNP laser, in which polarization dynamics can be adiabatically eliminated. The basic equations including a pump-modulation term are given by Eqs. (2.24)–(2.26) in

4.1. VARIATION OF LYAPUNOV EXPONENTS

which the term of injection-seeding $s_{i,k}$ is omitted. In the following, the population lifetime τ in Eqs. (2.24)–(2.26) is re-written as τ_e in order to distinguish it from the integral time τ in Eq. (4.5).

As is presented in **2.2**, the linear stability analysis of stationary states of Eqs. (2.24)–(2.26) proved that each mode possesses N relaxation oscillations $\omega_1 > \omega_2 > \cdots > \omega_N$, while the total intensity exhibits only one unique relaxation oscillation ω_1 just like a single-mode laser. If we assume a uniform gain distribution among lasing modes, i.e., $\gamma_k = 1$ independently of k, we can produce $(N-1)!$ antiphase "periodic" spiking oscillations. These appear as a repetitive generation of the first peak in the onset of relaxation oscillations, caused by a deep pump modulation, i.e., $\Delta w > w_0 - 1$, in a wide modulation frequency region below ω_1, i.e., $\omega_m = \omega_s < \omega_1$. When ω_m falls below ω_s, antiphase states are replaced by local chaos (clustered states) and finally global chaos showing chaotic spiking oscillations appears [see **2.2**].

Figure 4.4 shows temporal evolutions of photon densities in the chaotic

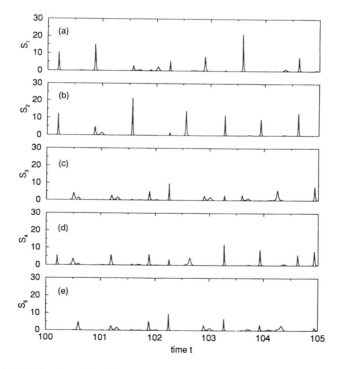

Figure 4.4: Simulated spiking laser intensities s_k in a five-mode laser, where uniform gain $\gamma_k = 1$, $w = 2.7$, $K = 1000$, $\epsilon = 1.2 \times 10^{-7}$, and $\tau_e \omega_m = 18.5$ are assumed.

regime, assuming $N = 5$, $w_0 = 2.7$, $K = 1000$, $\tau_e \omega_m = 18.5$, $\Delta w = 2$, and $\epsilon = 1.2 \times 10^{-7}$.

B. *Allan variance of local Lyapunov exponents variations*

Temporal evolutions of local Lyapunov exponents for the first (largest) and tenth Lyapunov exponents $\lambda_{1,10}(t)$ are shown in Fig. 4.5. Global Lyapunov spectra are shown in Fig. 4.6, where $\Lambda_4 = 0$ is due to the modulation term. Note that negative Lyapunov exponents are significantly larger than positive Lyapunov exponents of $O(1)$. This may be an inherent nature in spiking lasers. In addition, the temporal evolution of the largest local Lyapunov exponent well reflects the spiking nature of photon densities and the very narrow spikes appear for $\lambda_1(t)$ at peaks of spiking pulses as shown by the arrows in Fig. 4.5(a).

Figure 4.7 shows the standard deviation for $\lambda_{1,10}$ as a function of τ. The standard deviation becomes saturated in the short τ stage. This suggests that a variation of λ is continuous in time below a time scale of $\simeq 0.15$.

Allan variances of local Lyapunov spectra are calculated to attain an

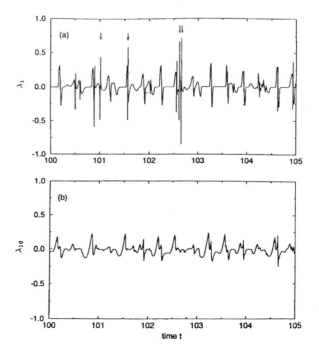

Figure 4.5: Examples of temporal evolutions of local Lyapunov exponents. (a) $\lambda_1(t)$, (b) λ_{10}. The adopted parameters are the same as those in Fig. 4.4.

4.1. VARIATION OF LYAPUNOV EXPONENTS

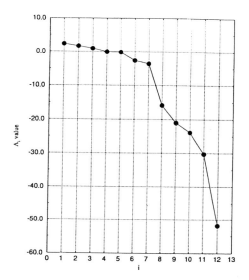

Figure 4.6: Global Lyapunov spectra. The adopted parameters are the same as those in Fig. 4.4.

understanding of the peculiar behavior of local Lyapunov exponents in short τ regimes. Examples for $\lambda_{1,10}$ are shown in Fig. 4.8. In this system, Allan variances in the large τ regimes exhibit strong undulation, resulting from the fundamental periodicity of pump modulation, i.e., $2\pi/\tau_e\omega_m$, as indicated by the arrows. This implies that the hidden modulation periodicity appears in local Lyapunov exponent variations, and is not seen in real variables s_k. In short τ regimes smaller than the relaxation oscillation period, i.e., $1/f_1 = 2\pi/\tau_e\omega_1$, Allan variances become saturated or even decrease with a scaling property. This strongly suggests a nonstationary nature of the variations, i.e., $S(f) \propto f^{-\alpha}$ ($\alpha \geq 1$) because Allan variance and power spectral density are connected by

$$\sigma^2(\tau) = 2\int_0^\infty S(f)\frac{\sin^4(\pi f\tau)}{(\pi f\tau)^2}df, \qquad (4.8)$$

and the mean reccurrence time becomes divergent when $\alpha \geq 1$, while in stationary chaos, i.e., $-\infty \leq \alpha < 1$, this reccurrence time becomes finite. Complex fluctuations without scaling properties appear in extremely small τ regimes. Similar Allan variances in nonstationary regime, with and without scaling properties, were reported for a modified one-dimensional Bernoulli map [3].

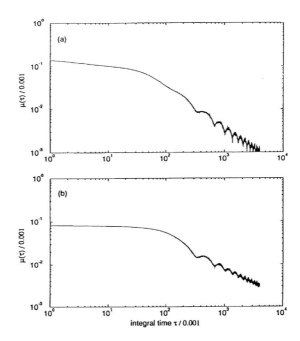

Figure 4.7: Standard deviations of local Lyapunov exponents (a) $\lambda_1(t)$ and (b) $\lambda_{10}(t)$ (b). The adopted parameters are the same as those in Fig. 4.4.

C. Power spectra of local Lyapunov exponents variations

Power spectra suggesting nonstationary chaos in short time stages with scaling properties were obtained by the fast Fourier transform (FFT) of local Lyapunov exponent variations. Examples of power spectra for the largest Lyapunov exponent λ_1 and a negative Lyapunov exponent λ_{10} are shown in Fig. 4.9. A very sharp noise corresponds to the modulation frequency $f_m = \tau_e \omega_m / 2\pi$, and $f_1 = \tau_e \omega_1 / 2\pi$ is the relaxation oscillation frequency. Dense bursts superimposed on the $S(f) \propto f^{-1}$ spectrum in high frequency regions for the first local Lyapunov exponent result from the very narrow (δ-function-type) spikes shown in Fig. 4.5(a). (See Fig. 4.9(a)). This causes complex fluctuations in extremely small τ regimes. These dense bursts are confirmed to be suppressed by coarse-graining (smoothing) of $\lambda(t)$ with an increase in average time Δt.

The power spectrum of $S(f) \propto f^{-5/3}$ is obtained for the tenth (negative) local Lyapunov exponent above the relaxation oscillation frequency f_1 as shown in Fig. 4.9(b). It is free of dense bursts because there are no narrow spikes in the negative local Lyapunov exponent variations as shown in Fig. 4.5(b). Below the relaxation oscillation frequency f_1, nonstationary

4.1. VARIATION OF LYAPUNOV EXPONENTS

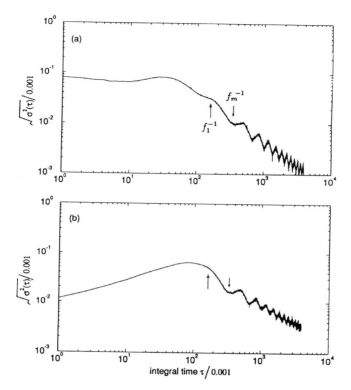

Figure 4.8: Square root of Allan variances for (a) $\lambda_1(t)$ and (b) $\lambda_{10}(t)$. The adopted parameters are the same as those in Fig. 4.4.

properties break down and the power spectrum of $S(f) \propto f^0$ appears, suggesting a Gaussian Markov property. This is also consistent with the scaling property of $\sigma_{10}^2 \propto \tau^{-1}$ in the large τ regime shown in Fig. 4.8(b). Such a nonstationary nature appears predominantly in negative local Lyapunov exponents like $\lambda_{9,10,11,12}(t)$ featuring $\alpha \geq 1$ above the relaxation oscillation frequency, in which α increases with i, while a Gaussian Markov property is recovered below the relaxation oscillation frequency independently of i. This suggests that significantly large negative local Lyapunov exponents control the present system, e.g., nonstationary properties. Intensive simulations for wide parameter regions indicate that the transition between stationary and nonstationary chaos occurs around the relaxation oscillation period independently of the system size, e.g., the number of oscillating modes.

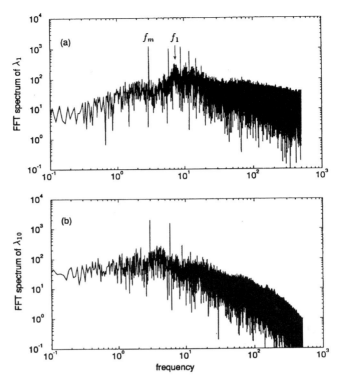

Figure 4.9: Power spectra of (a) $\lambda_1(t)$ and $\lambda_{10}(t)(t)$. The adopted parameters are the same as those in Fig. 4.4.

4.1.4 Summary and future problems

In summary, chaotic spiking oscillations in multimode lasers are characterized in terms of temporal variation of local Lyapunov exponents. Qualitatively different fluctuations resulting from stationary and nonstationary variations are separated by the time scale of relaxation oscillation period. The relationship between inverse power-law in intensity distribution in chaotic spiking lasers and the transition between stationary and nonstationary chaos is still undetermined and would be a future problem to be studied. Nonstationary fluctuation of local Lyapunov exponents in short-term chaotic dynamics below the relaxation oscillation period may, however, play an essential role. The inverse power-law relation and short-term memory in this chaotic system may also suggest the existence of some unique relation between the spiking periods (dead times) and the spiking peak intensities. In any case, the present system provides a realistic tool to use for investigating nonstationary chaos in strong connection with experiments.

4.2 Suppression of Chaotic Spiking Oscillations

Controlling chaos, which was initiated by Ott, Grebogi and Yorke [4], represents one of the most interesting programs in the field of nonlinear dynamics and has been been sucessfully demonstrated in various systems. The method is categorized into feedback-dependent and feedback-independent schemes. In optics, the control of laser chaos has been demonstated by the feeback-dependent OGY method [5] and by the modified OGY method [6]. The feedback-independent methods employing a weak periodic perturbation [7] or a weak parametric modulation [8] have been demonstrated in laser systems. However, these studies were restricted to a single-mode laser.

Another feedback-independent method, which uses the effect of near-resonant perturbation on subharmonic bifurcations which appear as precursors to chaotic behavior [9–12], has been experimentally demonstrated for controlling chaotic behavior of a magnetically driven magnetostrictive ribbon [13]. In this section, the near-resonant perturbation effect is described in a multimode laser system with the cross-saturation of population inversions (e.g., spatial hole-burning) and the suppression of multimode laser chaos is demonstrated. The key for successful demonstration is the use of a highly-sensitive multi-channel self-mixing laser-Doppler-velocimetry feedback scheme [see **3.3, 3.4**] for modulating a microchip solid-state laser.

4.2.1 Experimental results in a LiNdP$_4$O$_{12}$ microchip laser

A. Period-doubling bifurcation by single beam SMLDV modulation

The experimental arragement is shown in Fig. 4.10. The LiNdP$_4$O$_{12}$ (LNP) crystal is 1-mm-thick and both ends of the crystal are coated with dielectric mirrors (transmission at 1.32 μm of 0.1 and 1 percent). An argon laser serves as a pump. The TEM$_{00}$ mode oscillation threshold is $P_{th} = 130$ mW and the slope efficency is 13 percent. The lasing beam enters the two-channel Doppler-shifted light feedback controller through variable attenuators (VA), in which two beams were impinged on different rotating circular paper sheets. This is nothing other than the self-mixing laser-Doppler-velocimetry (SMLDV) scheme presented in **3.3, 3.4**. In this particular experiment, one modulation beam makes the laser chaotic, and the other is used for stabilization.

The first measurement is carried out by using single modulation beam. When the modulation frequency is chosen to the fundamental relaxation oscillation frequency of the laser $f_0 \equiv f_R = (1/2\pi)[(w-1)/\tau\tau_p]^{1/2}$ ($w =$

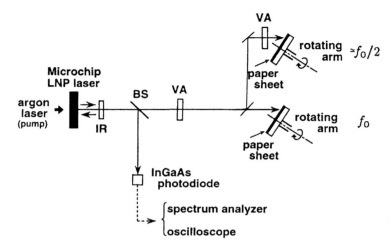

Figure 4.10: Experimental arrangement for simultaneous modulation at different frequencies.

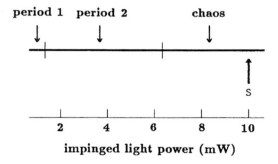

Figure 4.11: Bifurcation scenario of two-mode LNP laser as a function of light power impinged on the paper sheet in the absence of the second modulation beam. The modulation frequency is set to $f_0 = 420$ kHz. The LNP laser output power is 27 mW.

P/P_{th}: relative pump power), the total output displays a classic bifurcation from period 1 → period-2 → chaos for multimode oscillations, in which the entire set of lasing modes exhibits antiphase dynamics, as the feedback-light strength is increased [see **3.3**]. A brief bifurcation scenario for two-mode case is depicted in Fig. 4.11 as a function of a light power impinged on the paper sheet.

4.2. SUPPRESSION OF CHAOTIC SPIKING OSCILLATIONS

B. Suppression of chaos by contolling beam modulation

Next, the second 'controlling' beam path is open and and an additional weak modulation signal is fedback into the LNP laser. At this moment, each lasing mode is modulated by two feedback lights at different frequencies simultaneously [see **3.4**]. When the modulation frequency due to the second beam was set to be near-resonant to $f_0/2$, the unstable period-2 orbit embedded in a chaotic attractor is stabilized. Without the second controlling modulation beam, the system exhibits chaos as shown in Figs. 4.12(a)–(b). The broad peaks around f_0 and its higher harmonics are seen in the power

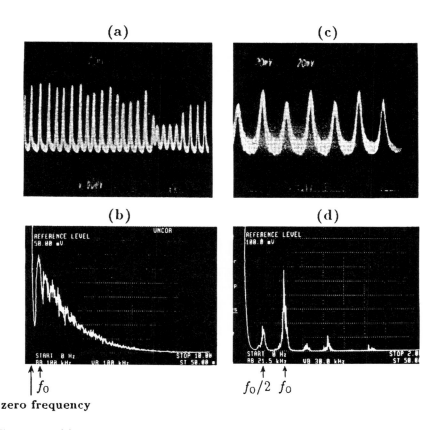

Figure 4.12: (a) Total output waveform showing chaotic relaxation oscillation when the second controlling beam is off. The impinged light power of the first modulating beam on the paper sheet was 10 mW (see point S indicate by an arrow). 5 μs/div. (b) Power spectrum for chaotic relaxation oscillation averaged over 1 second. 1 MHz/div. (c) Stabilized period-2 oscillation total ouput waveform when the second perturbation signal, which is near-resonant to $f_0/2$, is added at point S. The impinged light power on the paper sheet is 2 mW. 2 μs/div. (d) Corresponding power spectrum. 200 kHz/div.

spectrum. The stabilized period-2 oscillation waveform and the corresponding power spectrum are shown in Figs. 4.12(c)–(d), where the impinging light power of the second beam on the paper sheet is decreased to 1/5 that of the first beam. The sharp peaks at $f_0/2$, f_0, $3f_0/2$, $2f_0$ are clearly seen. These measurements are carried out at point S in Fig. 4.11 indicated by an arrow.

The suppression of chaos by the weak second near-resonant perturbation signal is very effective and the chaos in the entire region of Fig. 4.11 is completely replaced by the period-2 oscillation. The suppression of chaos may arise from a parametric amplification of the near-resonant $f_0/2$ subharmonic perturbation signal by the fundamental f_0 idler according to the theory of small signal amplification [12, 14]. The stabilization of period-4 orbit by near-resonant $f_0/4$ perturbation is not observed in the present parameter region because of the absence of period-4 bifurcation prior to

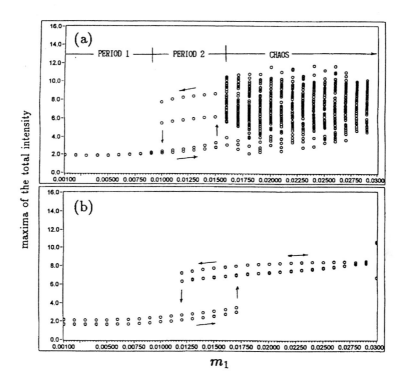

Figure 4.13: Calculated bifurcation diagrams with and without a controlling beam. (a) without controlling beam, (b) with a controlling beam, (c) Output waveforms with and without controlling beam.

4.2. SUPPRESSION OF CHAOTIC SPIKING OSCILLATIONS

transition to chaos in Fig. 4.11.

4.2.2 Numerical verification of suppressing chaos in a modulated multimode laser

These suppressing chaos experiments are well reproduced by numerical simulations of the following Tang-Statz-deMars (TSD) equations Eqs. (2.24)–(2.25) and (3.17) including multiple frequency loss-modulation term for globally coupled two-mode lasers with spatial hole-burning.

Bifurcation diagrams with and without a controlling beam are shown in Fig. 4.13, where $w = 2.6$, $K = 10^3$, $\gamma_1 = 1$, $\gamma_2 = 0.97$ and $\epsilon = 1.2 \times 10^{-7}$ are

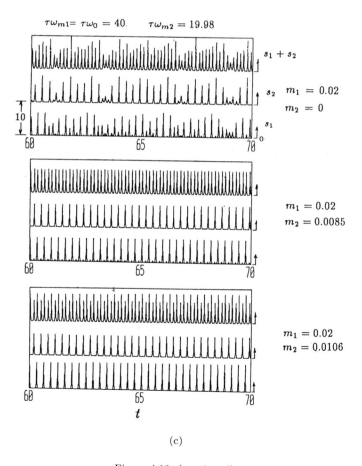

Figure 4.13: (continued).

assumed, in which the successive maxima of the total intensity $s_1 + s_2$ are plotted, except for transients, as a function of the modulation amplitude m_1. Figure 4.13(a) shows the bifurcation diagram without a controlling beam, where $\Omega_{m1} = \tau \omega_{m1}$ is chosen to the relaxation oscillation frequency $\Omega_0 = \tau \omega_0 = 40$. In this case, the system exhibits the bifurcation from period-1 → period-2 → chaos, featuring generalized bistability which is often observed in a loss-modulated class-B laser [15]. Here, the bistablity appearing for increasing and decreasing m_1 is indicated by arrows in the figure. The bifurcation diagram indicating the suppression of chaos by a perturbation at near-subharmonic resonance is shown in Fig. 4.13(b), where the controlling beam is added. Here, $\Omega_{m2} = \tau \omega_{m2} = 19.98$ and $m_2 = 0.0085$ are assumed. In this simulation, the chaotic region without the controlling beam is found to be completely suppressed by adding the controlling beam. Since the modulation amplitude m_i is proportional to $E_{i,s} = \sqrt{s_{i,s}}$ [see **3.3**, **3.4**], the "control" beam intensity ranges from 20 to 8 percent of the destabilizing beam intensity in Fig. 4.13(b). Examples of output waveforms with and without the controlling beam are shown in Fig. 4.13(c).

4.2.3 Summary and outlook

In summary, the effect of near-resonant subharmonic perturbation on chaotic dynamics is shown by using a microchip LNP solid-state laser with a highly-sensitive two-channel self-mixing laser Doppler velocimetry scheme. Chaos is first induced by the modulation at the fundamental relaxation oscillation frequency f_0 with scattered light from a rotating paper sheet which is reflected back into the michrochip LNP laser. The reflection from the second paper sheet is used to stabilize the chaotic dynamics. The experimental stabilization of the unstable period-2 orbit embedded in a chaotic attractor is performed by the application of the second weak perturbation signal which is near-resonant to a 1/2 harmonic of the fundamental frequency, $f_0/2$. The suppressing chaos experiments is well reproduced by numerical simulations of globally coupled multimode laser equations.

This system is *artificial* in the sense that chaos is first induced and then suppressed, however, the present near-resonant subharmonic perburbation method will find a useful application for stabilizing *autonomous* chaotic laser systems such as multimode solid-state laser with intracavity second-harmonic generation discussed in **2.3**, **2.4**, lasers with saturable absorber [7] etc. Dependences of a shift of the onset point of chaos on the control signal intensity and on the frequency detuning from the subharmonic resonance would be an interesting subject for future studies.

4.3 Chaotic Burst Generation in a Compound Cavity Multimode Laser

The complex dynamics and instabilities of lasers with external feedback have attracted much attention in the last decade. A semiconductor laser subjected to external feedback is discussed in Chapter **1**. The issue of chaotic instabilities in the output of lasers with external optical feedback was initiated by the pioneering work of Lang and Kobayashi in 1980 [16]. They demonstrated dynamical instabilities, featuring sustained relaxation oscillations, in a semiconductor laser with external feedback and confirmed theoretically, on the basis of a compound cavity laser model consisting of a laser diode cavity and an external cavity, that dynamical instabilities take place in the transition process where the lasing frequency changes from one compound cavity eigenmode to another. [See Figs. 1.17 and 1.18 in Chapter **1**.] Thereafter three universal transition routes to chaos, i.e., period doubling, quasiperiodicity, and intermittency, have been observed in semiconductor lasers with external optical feedback for different parameter regions as summarized in **1.4.3**. On the other hand, Otsuka observed chaotic burst generation in a cw LNP laser coupled to an optical fiber (multimode as well as single mode fibers) in 1979 [C-1, C-2].

Ikeda and Mizuno introduced a generalized model of compound cavity nonlinear passive resonators [17–18] and lasers [19]. They showed theoretically that a competition between two time-delayed feedbacks causes "frustration" in selecting an oscillating mode, when there exist many potential oscillating modes with subtly different stabilities [17–18]. The mechanism is similar to the frustration phenomena in thermal equilibrium systems [20], resulting from competing interactions of order parameters. They predicted, on the basis of a number-theoretic method, that a slight change in cavity length enables an oscillation in a quite different mode.

In this section, periodic output intensity drops are shown in the LNP laser coupled to a single-mode optical fiber as the phase shift within the LNP cavity is swept in time. Chaotic output power drops, evidence of "frustration" in selecting an oscillating mode, are shown to occur when the sweep rate is increased in the pump regime below the solitary LNP laser threshold $P_{th,s}$, i.e., $P_{th,c} < P \leq P_{th,s}$, where $P_{th,c}$ is the compound-cavity laser threshold. Chaotic relaxation oscillation bursts, reflecting dynamical instabilities, is shown to appear corresponding to these drops as the pump power is increased above the solitary LNP laser threshold $P_{th,s}$. A random generation of chaotic bursts, which results from a mode-partition-noise-induced random phase shift, also takes place in multimode oscillation regimes above $P_{th,s}$ without any intentional phase sweeping.

4.3.1 Frustration phenomenon: periodic and chaotic intensity drop

A. Experimental setup

The experimental apparatus is shown in Fig. 4.14(a). The LNP crystal used in the experiment is the same as previous sections. A highly stable argon laser light being stabilized by an external optoelectronic feedback device ("noise eater") is employed as a pump source to avoid pump fluctuations, where the fluctuation amplitude is less than 0.1 percent. The oscillation threshold is 130 mW and the slope efficiency is 13 percent. The TEM$_{00}$ 1.32 μm LNP laser radiation is linearly polarized along the pseudoorthorhombic b axis. The LNP cavity mode spacing is $\Delta\lambda = 0.48$ nm, where the gain spectrum half-width is 3 nm. Part of the output light is used for monitoring output waveforms. The main beam is tightly coupled to a 10m-long single mode optical fiber (core diameter: 7.5 μm, loss at 1.32 μm: 1 dB/km) by a microscope objective lens, where the lasing threshold $P_{th,c}$ is decreased by 5 percent as compared with the solitary LNP laser threshold $P_{th,s}$.

Here, the delay time of $t_D \simeq 100$ ns is much smaller than the coherence time and the population lifetime of $\tau = 120$ μs of the LNP laser. The LNP laser cavity together with the polished fiber-end reflector having a 4 percent reflectivity forms a compound cavity laser system. Indeed, the following

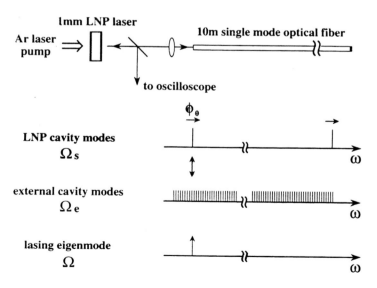

Figure 4.14: Experimental setup and frequency arrangements of the LNP laser coupled to a single-mode optical fiber.

4.3. CHAOTIC BURST GENERATION

instabilities are completely suppressed by dipping the fiber end in matching oil. The external cavity mode spacing is 10 MHz and 10^4 external cavity modes exist between the LNP cavity modes. A lasing eigenmode frequency of the compound cavity, Ω, is determined from the frequency arrangement of the LNP laser cavity modes $[\Omega_s] = [\omega_{s,1}, \omega_{s,2}, \ldots, \omega_{s,n}]$ and external cavity modes $[\Omega_e] = [\omega_{e,1}, \omega_{e,2}, \ldots, \omega_{e,N}]$ ($N \gg n$) on the basis of the field continuum condition at output mirror [16]. The frequency arrangement of the system is depicted in Fig. 4.14(b).

B. Intentional phase sweep

First, let us examine the system response to the phase sweep in the LNP cavity in time. For this purpose, the turn-on characteristic is utilized by chopping the pump power with a mechanical shutter. Figure 4.15(a) shows LNP laser output power change without the optical fiber, where the time scale is 0.5 sec/div. The output power decreases exponentially as a result of the thermal effect and approaches the stationary state.

Together with the thermally-induced refractive index change, the output power change results in the LNP refractive index change because of the dependence of refractive index on population inversion density in the LNP laser. In microcavity LNP lasers, such an anomalous dispersion effect is easily brought about by detuning the lasing frequency from the gain spectrum peak. It is done by shifting the pump position within the LNP crystal which has a slight position dependent thickness. The phase sweep rate of the LNP laser cavity resonance is given by $V \equiv d\phi_0/dt = (4\pi n_2(\Delta\omega)\ell/\lambda)(dI/dt)$, where $n_2(\Delta\omega)$ is the nonlinear refractive index resulting from the anomalous dispersion effect ($\Delta\omega$: detuning frequency), ℓ is the LNP length, and I is the laser intensity in the LNP cavity. Therefore, from Fig. 4.15(a) the phase sweep rate is found to gradually decrease as time develops.

Figures 4.15(b)–(d) show waveforms observed at different time frames after the turn-on, where the pump power is chosen *in the vicinity of the solitary LNP laser lasing threshold*, i.e., $P \simeq P_{th,s}$. Near the stationary state (b), the sweep rate is low and periodic drops of laser intensity are clearly seen, where the sweep rate changes slightly within the observation time scale of 2 msec/div. When the time approaches the turn-on time, i.e., sweep rate increases, the periodicity fails and finally the intensity drop tends to occur chaotically, as shown in (d). Fast fluctuation components correspond to intrinsic relaxation oscillations around 100 kHz. They are driven by a white noise in the system as described in Chapter **3** and are not essential.

Exactly the same waveforms are predicted by numerical simulations and the intensity drops are shown to occur in transition regimes $\delta\omega_u$ between stable compound cavity lasing eigenmodes [19], where a stable single frequency oscillation is impossible. Numerical examples at different sweep

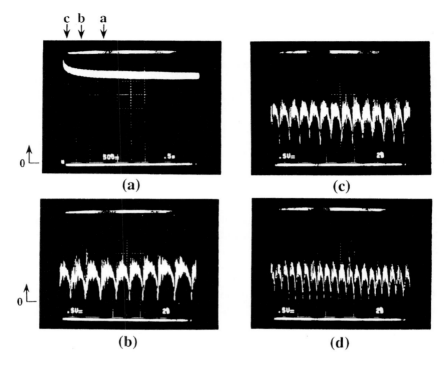

Figure 4.15: (a) Turn-on characteristic of the solitary LNP laser output without the optical fiber when the pump power is increased in a stepwise fashion by a mechanical chopper. 0.5 sec/div. (b)–(d) Periodic and chaotic drops of the LNP laser output intensity coupled to the optical fiber observed at different times a, b, and c after the turn-on. The pump power is chosen in the vicinity of the solitary LNP laser threshold, i.e., $w = P/P_{th,c} = 1.05$, where $P_{th,c}$ is the compound cavity laser lasing threshold. 2 msec/div. (b) waveform at time a, (c) at b, and (d) at c.

velocities are shown in Fig. 4.16 [19]. In short, it is considered that observed intensity drops take place when the eigenmode frequency Ω enters these regimes where the solitary LNP cavity mode field is almost absent, which exist between different stable eigenmode states. In particular, chaotic intensity drops at higher sweep rate (i.e., large phase shift in time) correspond well to dynamics resulting from the predicted frustration phenomena [19].

Note that without an intentional phase sweep, the subharmonic resonance cascade leading to chaotic relaxation oscillations observed below $P_{th,s}$ in an LD strongly coupled to an external cavity [see Fig. 1.21] does not occur in the present system. Presumably, this is due to the fact that amplified spontaneous emission (ASE) mode is extremely weak in LNP

4.3. CHAOTIC BURST GENERATION

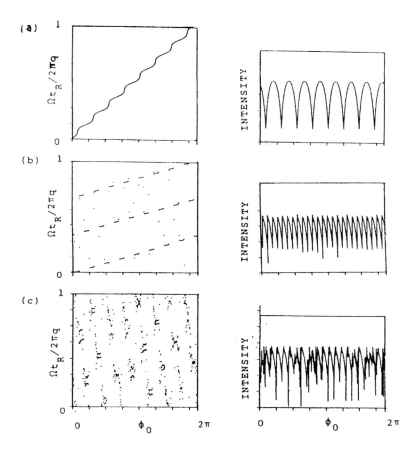

Figure 4.16: Numerical results showing variation of eigenmode frequency Ω and lasing intensity when the phase shift is varied with different scanning velocities.

lasers as compared with a high gain laser diode. Consequently, parametric four-wave mixing process [see Fig. 1.22] is considered to be weak enough to bring about autonomous instability.

When the pump power is increased *above the solitary LNP laser threshold* $P_{th,s}$, chaotic relaxation oscillations appear at times when intensity drops (Figs. 4.15(b)–(d)) occur. The results are shown in Fig. 4.17 at different times after the turn-on. A periodic generation of chaotic bursts is apparent at low sweep rates (a) and (b), while a chaotic generation occurs at high sweep rates (c). A similar instability featuring relaxation oscillations is demonstrated in "transition regimes" from one stable compound cavity lasing eigenmode to another in a laser diode with external optical

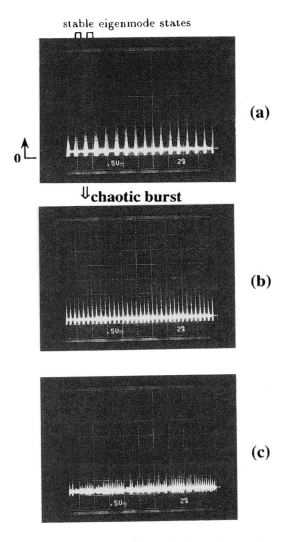

Figure 4.17: Periodic and chaotic generations of chaotic bursts observed at different times after the turn-on. Pump is increased above the solitary LNP laser threshold, e.g., $w = 2$. 2 msec/div. (a) 1.5 seconds after the turn on, (b) 0.8 seconds, and (c) 0.1 seconds. Stable eigenmode states are indicated by ⊓. Chaotic bursts appear between different eigenmode states.

feedback [16], in which sustained relaxation oscillations occur at "undulation bottoms" in the I-L curve [see Figs. 1.17 and 1.18]. Such an instability is derived from the nonlinear interaction between population inversion and

4.3. CHAOTIC BURST GENERATION

light field, i.e., Eq. (1.44). Dynamical instabilities occur when the lasing eigenmode frequency Ω satisfies the condition in the limit $\zeta t_d \ll 1$ [16]:

$$\zeta t_d[-\alpha \sin(\Omega t_D) + \cos(\Omega t_D)] + 1 < 0, \qquad (4.9)$$

where ζ is the coupling coefficient, t_D is the delay time and $\alpha = 4\pi n_2 I_s/\lambda g_0$ (I_s: saturation intensity, g_0: small signal gain coefficient). For the LNP, $|\alpha| = O(1)$. An example of numerical results based on the compound cavity laser model [16], [21] is shown in Fig. 4.18 under the condition of constant phase sweep.

Coherence collapse instability above $P_{th,s}$ observed in an LD strongly coupled to an external cavity is not occurring in the present system.

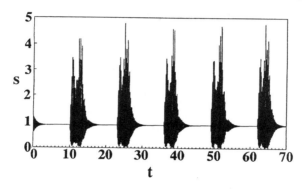

Figure 4.18: A numerical result showing the periodic generation of chaotic bursts under a constant phase sweep condition at $w = 2$ and $\zeta t_D = 1.67$. s is the normalized output intensity and the time t is scaled by the population lifetime τ.

4.3.2 Mode-partition-noise-induced chaotic bursts

It is easily expected that the state of frustration resulting from a slight change in the laser cavity length occurs because of the small 10 MHz free spectral range of the external cavity modes. To check this idea, let us examine output waveforms in stationary states *without intentional phase sweeping*.

In the regime where the solitary LNP laser without the fiber exhibits single longitudinal mode oscillations, no instability takes place. This implies that the output intensity of the solitary LNP laser is quite stable and the phase shift is too small to bring about lasing frequency transitions. If the pump power is increased up to regimes where the solitary LNP laser exhibits multimode oscillations, the mode-partition-noise that has a *quantum mechanical origin* appears, in which the total output is partitioned

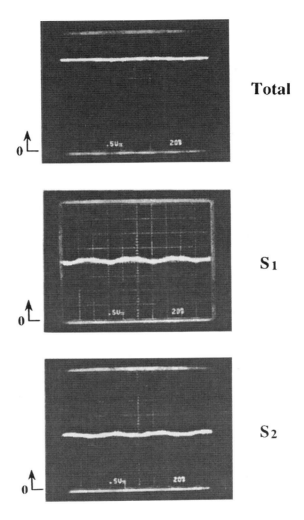

Figure 4.19: The total output, $s_1 + s_2$, and the modal output waveforms, s_1, s_2, of the solitary LNP laser oscillating in two longitudinal modes. The modes are selected by using a spectrometer. $w = 2$. 20 msec/div.

into different modes at random with shot noise and individual modes show low-frequency fluctuations whose noise level increases with decreasing frequency, while the total output is free from these fluctuations [22]. Such an instrinsic mode-partition-noise exists in multimode lasers in addition to multiple antiphased relaxation oscillation noise that has a *dynamical ori-*

4.3. CHAOTIC BURST GENERATION

gin described in **3.2**. The modal and total output intensities are shown in Fig. 4.18.

In ordinary solid-state lasers like LNP, multimode oscillations result from the spatial hole-burning effect of population inversion, in which the gain saturation is mode-dependent [23]. Consequently, together with the

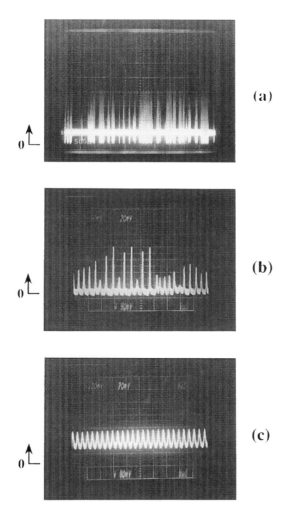

Figure 4.20: (a) Random chaotic bursts observed in multimode regimes. 20 msec/div. (b) Waveform of (a) on the extended time scale. 5 μsec/div. (c) Sustained relaxation oscillation burst with a decreased coupling efficiency. 5 μsec/div.

frequency dependent nonlinear refractive index n_2, the LNP refractive index change becomes mode-dependent. As a result, the random intensity fluctuation in each mode results in the mode-dependent random fluctuation in phase shift, i.e., $[\Omega_s]$. In this situation, chaotic bursts are expected to occur at random in the presence of an external cavity, when the LNP cavity mode frequency fluctuates and the lasing eigenmode frequency Ω enters dynamically unstable zones $\delta\omega_u$. As expected, chaotic bursts appears as shown in Fig. 4.20(a) even in the absence of intentional phase sweep. Figure 4.20(b) is the detailed waveform of chaotic burst. The oscillation frequency coincides with the relaxation oscillation frequency. When the coupling efficiency into the fiber (ζ) is decreased, random generation of sustained relaxation oscillation bursts is observed as shown in Fig. 4.20(c). Here, each burst consists of periodic sustained relaxation oscillations just like that observed at undulation bottoms in an LD weakly coupled to an external cavity [see Fig. 1.17].

These observations indicate that frustrated instabilities featuring chaotic relaxation oscillation bursts generation take place in multimode lasers coupled to an external cavity as a result of the intrinsic mode-partition-noise without intentional phase sweeping. In other words, observed chaotic bursts can be regarded as mode-partition-noise-mediated dynamical instabilities. A conceptual model of the chaotic burst generation process is illustrated in Fig. 4.21.

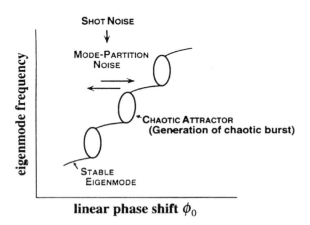

Figure 4.21: Conceptual illustration of chaotic burst generation in a multimode coumpound-cavity laser.

4.3.3 Summary and outlook

In this section, periodic and chaotic drops of output intensity accompanied by chaotic relaxation oscillation bursts in a microchip solid-state laser coupled to an optical fiber are presented under the constant phase sweeping, which strongly suggests the frustration phenomena in a compound cavity laser system.

The intrinsic mode-partition-noise-mediated dynamical instability featuring chaotic bursts is shown in multimode regimes without intentional phase sweeping. This observed random chaotic burst generation originated from quantum noise (e.g., shot noise) will provide a new insight into multimode laser instabilities. In short, the present system features a *double structure* of "quantum" noise and "dynamical" instabilities and the quantum fluctuation is manifested in the form of chaotic bursts.

4.4 Breakup of CW Multimode Oscillations by High-Density Pumping

4.4.1 Introduction

The spatial hole-burning effect of population inversion resulting from lasing standing-wave patterns in Fabry-Perot laser resonators was introduced by Tang, Statz and deMars (TSD) in early 60's [24]. It is the fundamental physical process which allows multimode oscillations in homogeneously-broadened solid-state lasers. This is due to the slow spatial diffusion of population inversions in solid-state laser crystals such as YAG. TSD expanded the distribution of population inversion into spatial Fourier components and derived the so-called TSD equations by retaining only the space average and Fourier components varying at optical wavenumbers [see Eqs. (3.1)–(3.3)]. The relation between lasing mode number and pump rate was deduced from the TSD equations and was verified experimentally in various solid-state lasers including Nd:YAG lasers [25].

The recent rapid progress of high power laser-diode-pumped microchip lasers has driven a renaissance of solid-state laser physics research and led to novel phenomena, such as winner-takes-all dynamics (e.g., 'pony on the merry-go-round') and antiphase periodic states, antiphase dynamics, dynamical nonreciprocal independence and universal power spectra relation in multimode lasers, which are described in Chapters **2** and **3**. One of the most peculiar features inherent in high-power laser-diode pumped miniature microchip solid-state lasers is the Auger recombination process among excited ions, which is expected in microchip lasers with high-density pumping [26]. In the Auger process, two excited electrons in the $^4F_{3/2}$ level interact due to excited state absorption: one electron makes a transition to the lower laser level $^4I_{11/2}$ emitting a 1.05μm photon, while another electron is excited to the higher energy level $^2GI_{9/2}$. The Auger coefficient is proportional to excited ion density N_e and a strong Auger effect is expected in state-of-art high-power laser-diode-pumped microchip lasers with high density pumping such as Nd:YVO, Nd:LSB, Cr:LiSAF etc. Particularly, Nd stoichiometric (e.g., direct compound) lasers such as NdP$_5$O$_{14}$, LiNdP$_4$O$_{12}$, NdAl$_3$(BO$_3$)O$_4$ and Nd:CeCl$_3$ developed in early 70's, in which the Nd ion concentration is 30 times higher than in conventional microchip lasers, are promising candidates for investigating the effect of Auger recombination process on spatial hole-burning and multimode oscillations.

In this section, the effect of high-density pumping on spatial hole-burning is investigated experimentally and theoretically. First, experimental results concerning the effect of high-density pumping on lasing mode spectra are shown by using a laser-diode pumped microchip LNP laser. It is shown that the breakup of multimode oscillations, leading to single-mode oscillation, takes place above a critical pump density threshold. Next, analytical results on the high-density pumping effect are described and it is shown that the Auger recombination process inhibits the multimode regime in favor of the single mode regime. A simple criterion is derived for the critical pump power at which the multimode operation becomes unstable. These theoretical results are shown to reproduce experimental results quite well. Finally, a puzzling mode-partition instability associated with multiple transition oscillations in a Λ-scheme is shown experimentally.

4.4.2 Experimental results

A. Input-output characteristics

Experiments are carried out by using a laser diode (LD)-pumped LNP laser. An experimental setup of an LD-pumped microchip LNP laser is depicted in Fig. 4.22. The LNP crystal is 1-mm thick and dielectric mirrors are coated on both ends of the crystal. The output coupling is 1 percent in 1000–1100 nm range. A collimated lasing beam from the laser diode oscillating at $\lambda_p = 808$ nm is passed through anamorphic prism pairs to transform an elliptical beam into a circular beam and is focused onto the LNP crystal by a microscopic objective lens ($M \times 40$). By changing the distance between the lens and the LNP crystal, the pump beam spot size is controlled. A linearly-polarized TEM$_{00}$ mode oscillation along the psuedoorthorohmbic b-axis is observed.

Input-output characteristics measured for different pump spot sizes are shown in Fig. 4.23. Here, the effective pump beam spot size w_p averaged over the absorption length $\ell_p \simeq 1/\alpha_p = 100$ μm ($\alpha_p \simeq 100$ cm^{-1} is the absorption coefficient of LNP at $\lambda_p = 808$ nm) is estimated to be 50 μm (a), 40 μm (b) and 25 μm (c) from the pumped cross section, respectively. Such

4.4. BREAKUP OF CW MULTIMODE OSCILLATIONS

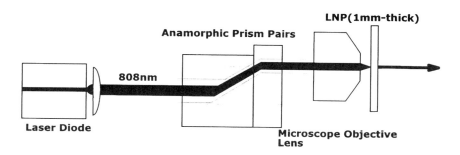

Figure 4.22: Experimental apparatus of a laser-diode pumped LNP laser.

a short absorption length implies that the population inversion is strongly localized at the input mirror region inside the laser resonator. The number of oscillating modes is added in the figures. The oscillating beam spot size is estimated to be $w_o \simeq 150$ μm and the slope efficiency is seen to increase as the pump spot size w_p is decreased.

In the case of Fig. 4.23(a), we see from the measured threshold pump power that $N_e/N_0 = N_{th}/N_0 \simeq 0.028$, where N_{th} is the threshold Nd density for lasing, N_e is the excited Nd density and N_0 is the Nd concentration in LNP crystals [27]. In this case, usual input-output characteristics, in which the number of oscillating modes increases with the pump power [24, 25] are obtained. When the pump power density is increased by decreasing the pump beam spot size, unusual input-output characteristics are observed as shown in Fig. 4.23(b). In the case of Fig. 4.23(b), $N_e/N_0 \simeq 0.046$ and the breakup of multimode oscillation is occurring at $w \equiv P/P_{th} = w_c \simeq 2$ indicated by an arrow, at which a number of modes begins to decrease, leading to single-mode oscillation, where P is the pump power and P_{th} is the threshold pump power. As the pump power density is increased further like in Fig. 4.23(c), spontaneous single-mode oscillation is observed in the entire pumping domain. In the case of Fig. 4.23(c), the extremely high density pumping of $N_e/N_0 \simeq 0.1$ is attained. In fact, in LNP crytals, each Nd ion has 8 nearest neighbor Nd ions [28]. Therefore, in the case of Fig. 4.23(c), an excited ion will, on average, begin to see one excited ion among its nearest neighbors, assuming $N_0 = 4.37 \times 10^{21}$ cm^{-3} [27].

Observed spontaneous single-mode oscillations are very attractive from the view point of practical applications. In particular, single-mode oscillations in a 1-mm-thick LNP laser ensures laser-diode pumped single-frequency intracavity second-harmonic generations (ISHG) by attaching a frequency-doubling crystal to the LNP crystal. The single-mode oscillation in ISHG enables us to avoid instabilities (e.g., "green problem") resulting from ISHG in multimode lasers described in **2.3**, **2.4**.

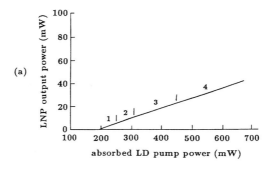
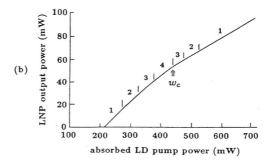
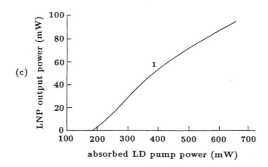

Figure 4.23: Input-output characteristics and number of oscillating modes for different pump spot sizes (e.g., densities).

B. *Multiple-transition oscillations and antiphase dynamics*

In the case of multimode class-B lasers with spatial hole-burning, in which polarization lifetime is much shorter than the population inversion lifetime τ, it has been established from intensive theoretical and experi-

4.4. BREAKUP OF CW MULTIMODE OSCILLATIONS

mental studies that the total output features damped relaxation oscillations at a unique frequency $f_1 = f_R = (1/2\pi)\sqrt{(w-1)/\tau\tau_p}$ (τ_p: photon lifetime), while each mode exhibits N relaxation oscillations, namely $f_1 > f_2 > \cdots > f_N$, where N is a number of lasing modes [see Chapter 3]. The so-called McCumber frequency f_1 corresponds to the relaxation oscillation frequency in a single mode laser derived from the simple linear stability analysis of a single-mode laser. In other terms, individual modes are self-organized to exhibit antiphase dynamics such that low frequency relaxation oscillations are strongly suppressed in the total output.

In the present experiment, such self-organized nature concerning antiphase relaxation oscillations is confirmed in multimode oscillation regimes independently of the pump power density. Examples are shown in Fig. 2.24, where measurements are carried out in multimode regimes below the critical pump w_c in Fig. 4.23(b). Left-hand photographs correspond to power spectra for the first lasing mode at the wavelength $\lambda_{\ell,1} = 1047.66$ nm, which corresponds to the $^4F_{3/2}(1) \rightarrow\ ^4I_{11/2}(1)$ transition, and right-hand photographs correspond to power spectra for the total output. It is apparent that relaxation oscillation peaks are strongly suppressed in the total output, except at the McCumber frequency f_1 which are driven by a 'white' noise. We shall refer to this property later on in this section as generic antiphase property. In Fig. 4.24, the second lasing mode appears via the $^4F_{3/2}(1) \rightarrow\ ^4I_{11/2}(2)$ transition at $\lambda_{\ell,2} = 1055$ nm, the third lasing mode appears via the $^4F_{3/2}(1) \rightarrow\ ^4I_{11/2}(3)$ transition at $\lambda_{\ell,3} = 1060$ nm and the fourth lasing mode appears at $\lambda_{\ell,4} = 1048$ nm which belongs to the $^4F_{3/2}(1) \rightarrow\ ^4I_{11/2}(1)$ transition. Note that the longitudinal mode spacing is an order of magnitude larger than the value determined by the cavity length ℓ, which gives $\Delta\lambda = \lambda^2/2n\ell = 0.34$ nm. On the contrary, above the critical pump w_c, the 1048 nm mode stops lasing at first as the pump power is increased, the 1047.66 nm mode is turned off next, and then the 1055 nm mode is turned off, and finally single-mode oscillation at 1060 nm survives.

Similar multiple-transition oscillations featuring "cross mode" (see Fig. 10(c) of [29]) is reported in Ar laser pumped LNP lasers reflecting the decrease in reabsorption loss of the LNP crystal, $L_{ra}(1048) > L_{ra}(1055) > L_{ra}(1060)$. Each of these transitions supports the lasing of many longitudinal modes separated by $\Delta\lambda$. In the present case, on the contrary, due to the strong localization of the population inversion near the input mirror, nearby modes separated by $\Delta\lambda$ are shown to be strongly quenched as a result of the Lamb's mode competition through the strong cross saturation [D-2], [23] in favor of sparse mode oscillations belonging to different transition lines. Generic antiphase relaxation oscillation property is observed both below and above the critical pump. From these observations, it is concluded that antiphase dynamics inherent in multimode lasers with spa-

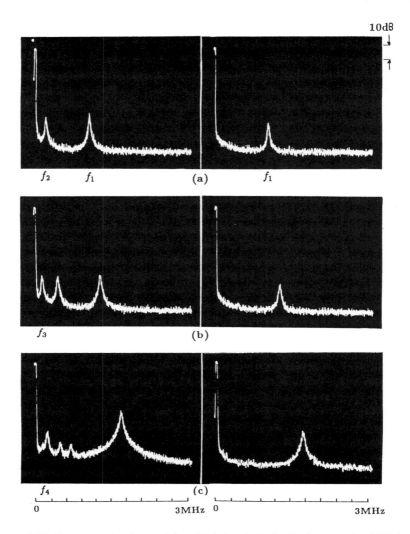

Figure 4.24: Power spectra for modal and total outputs in the free-running LNP laser corresponding to Fig. 4.23(b). (a) Two-mode regime, (b) three-mode regime and (c) four-mode regime. Left-hand figures indicate power spectra of the first lasing mode and right-hand figures indicate power spectra for the total output. For the total output, input power to rf spectrum analyzer was attenuated by -10 dB. vertical scale: 10 dB/div. horizontal scale: 300 kHz/div.

tial hole-burning holds even in the case of *multiple-transition oscillations* by localized high-density pumping.

C. Ar laser and Kr laser pumping

Experimental results described so far strongly implies that high-density pumping results in unusual input-output characteristics. To verify this idea more precisely, other pump sources with different absorption coefficients α_p are used instead of the laser diode, to change pump power density more drastically. For this purpose, an Ar laser ($\lambda_p = 5145$ A, $\alpha_p = 23$ cm^{-1}) or a Kr laser ($\lambda_p = 7993$ A, $\alpha_p = 250$ cm^{-1}) served as a pump. The absorption lengths $\ell_p \simeq \alpha_p^{-1}$ are 435 μm for 5145 A and 40 μm for 7993 A, respectively.

In the case of Ar laser pumping, in which absorption length is much longer than LD pumping, usual multimode oscillation characteristics like Fig. 4.23(a) are observed for various pump beam spot sizes. For Kr laser pumping, in which the pump wavelength matches with the absorption peak of LNP crystals, spontaneous single-mode oscillations are established in entire pump regimes.

4.4.3 Theoretical results

A. Smoothing effects of spatial hole-burning pattern

The favored single-mode oscillations in high pump density regimes (e.g., Figs. 4.23(b) and (c)), strongly suggests that spatial hole-burning patterns tend to be smeared out as the pump power density is increased. A first source of smoothing of spatial hole-burning pattern is resonant transfer, i.e., reabsorption process [27], as shown in Fig. 2.25(a). Since the relaxation energy is transferred in full to one of the neighbor *nonexcited* ions, all fluorescence transitions can contribute to resonance transfer. This is effective since the overlapping between emission and absorption lines is perfect. Especially, the $^4I_{9/2}(1) \rightarrow\, ^4F_{3/2}(1)$ transition from the ground state is the most effective in resonance transfer for LNP crystals. Based on the spectroscopic data of LNP [27], the transfer probability of excitation energy to one of the neighbor ions via the $^4I_{9/2} \rightarrow\, ^4F_{3/2}$ reabsorption process is estimated to be $W_J = 1.2 \times 10^8$ s^{-1}, which yields the diffusion distance $d\sqrt{W_J\tau} = 800$ Å in LNP crystals, where $d = 6$ Å is the mean separation between Nd ions and $\tau = 120$ μs. Since this is much shorter than a quarter-wave length, spatial hole-burning pattern is not sufficiently smeared out by this energy migration process in LNP crystals.

The second possible mechanism is Auger recombination process depicted in Fig. 4.25(b), in which two *excited* Nd ions interact and the pair annihilation takes place through excited state absorption via the $^4F_{3/2} \rightarrow\, ^2GI_{9/2}$ transition in LNP crystals. Since this process occurs in proportion to N_e^2, the enhanced deformation of spatial hole-burning pattern is expected in high pump density regimes. In the following, the effect of Auger recombination process on spatial hole burning pattern is analyzed theoretically.

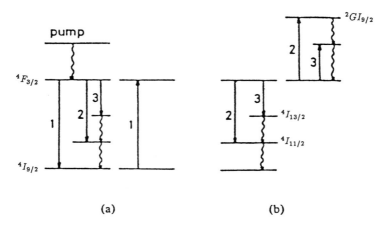

Figure 4.25: Models of (a) fluorescence reabsorption process and (b) Auger recombination process in LNP lasers. Here, **1** indicates the transition between the ground state and the excited state at 870 nm, **2** corresponds to the lasing transition at 1048 nm for ↓ and corresponding excited state absorption for ↑, **3** corresponds to the lasing transition at 1320 nm for ↓ and corresponding excited state absorption for ↑.

B. *Quadratic-to-quartic transition of spatial hole-burning pattern*

The rate equation for the population inversion density $N(z)$ including Auger recombination process is given by

$$\frac{dN(z)}{dt} = P(z) - \frac{[1 + q_2 N(z)]N(z)}{\tau} - B_c N(z) S(z), \quad (4.10)$$

where $P(z)$ is the pump rate, τ is the population lifetime, q_2 is the Auger parameter, B_c is the stimulated emission coefficient, and $S(z)$ is the photon density. Let us introduce the population inversion density at the lasing threshold averaged over the absorption length $N_{th} = (1/\ell_p) \int_0^{\ell_p} N(z) dz$. For the LNP laser, $\ell_p \simeq 100$ μm. Defining a normalized population inversion through $n(z, t) \equiv N(z, t)/N_{th}$, we find in steady state

$$n(z) = \frac{-1 - s(z) + \sqrt{[1 + s(z)]^2 + 4q_2 N_{th}(z) w(z)}}{2Q_2}. \quad (4.11)$$

Here, $w = \tau P/N_{th}$ is the normalized pump power, $s(z) = B_c S(z) \tau$ is the normalized photon density and $Q_2 \equiv q_2 N_{th}$ is the normalized Auger parameter. Assuming $4q_2 N_{th} w \ll [1 + s(z)]^2$ and $s(z) = s_1 \sin^2 k_1 z$, where k_1 and s_1 are, respectively, the wavenumber and the photon density of the first lasing mode, Eq. (4.11) is approximated by

$$n(z) \simeq \frac{w}{1 + s_1 \sin^2 kz} - \frac{Q_2 w^2}{[1 + s_1 \sin^2 kz]^3}, \quad (4.12)$$

4.4. BREAKUP OF CW MULTIMODE OSCILLATIONS

where we have also assumed $w = (1/\ell_p) \int_0^{\ell_p} w(z)dz$. In NdP_5O_{14} lasers, for example, $Q_2 = 1.8$ when the excited Nd ion density N_e is 50 percent of the total Nd ion concentration N_0 [26]. Equation (4.12) expresses the population inversion distribution when only the first lasing mode is oscillating.

If Q_2 is sufficiently small, the function $n(z)$ has two maxima and two minima in a period $0 \leq kz < 2\pi$. If Q_2 is sufficiently large, the function $n(z)$ has four maxima and four minima in each period. At the transition between these two domains, two extrema of $n(z)$ coincide which means that $n(z)$ has quartic instead of quadratic extrema. Expanding $n(z)$ around its maximum $n(z) = n(z_{\max}) + (z - z_{\max})^2 n'' + \cdots$, the quadratic-to-quartic (Q2Q) transition is characterized by $n'' = 0$. Solving this equation for the critical pump at which this transition occurs yields the relation

$$w_c = \frac{1}{3Q_2}. \qquad (4.13)$$

This simple *univeral relation* implies that *the transition occurs independently of the first lasing mode intensity s_1 and that the critical pump is determined only by the Auger parameter.*

A numerical example based on Eq. (4.12) is given in Fig. 4.26. Here, $Q_2 = 0.1, 0.17$ and 0.36 are assumed, where these values correspond to the estimated Q_2 values for Figs. 4.23(a), (b) and (c), respectively. For moderate values of Q_2, as in Fig. 4.26(a), the quadratic-to-quartic transition is not reached for reasonable values of the pump rate. For larger Q_2, the transition is reached for finite pumping ($w_c \simeq 2$ for $Q_2 = 0.17$) and the population inversion distribution is strongly deformed above w_c as shown in Fig. 4.26(b). The w_c-value agrees with the analytic result obtained from Eq. (4.12) as well as with the observed value in Fig. 4.23(b). Consequently, the overlap integral between empty cavity eigenmodes and the deformed population inversion pattern is decreased as the pump is increased in the regime above w_c. Eventually, multimode oscillations becomes unstable. This result parallels the change of slope observed around w_c in Fig. 4.23(b). At the critical pump $w = w_c$, the population distribution displays the plateau characteritic of the quartic maximum as shown in Fig. 4.26(b). Finally, in the case $Q_2 \geq 1/3$, shown in Fig. 4.26(c), the standing-wave pattern is no more sinusoidal right from the lasing threshold, leading to spontaneous single-mode oscillation. In other terms, the dip in the center of $n(z)$ corresponds to the node of the eigenfuctions, whereas side dips cut excess energy in antinodes.

Although the condition given by Eq. (4.13) is derived in the single mode regime, it is verified numerically that it still gives a good approximation of w_c in the multimode regime in the range of pump rate used in the experiments, in which a number of modes is not large.

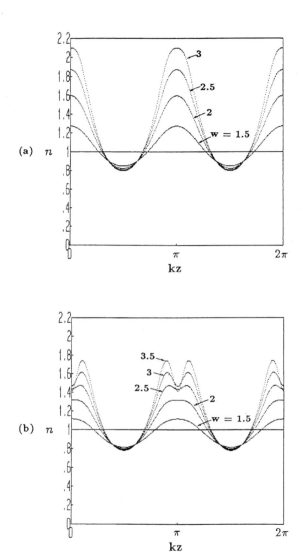

Figure 4.26: Numerical spatial distribution of population inversion for various pump levels when the first lasing mode is oscillating. (a) $Q_2 = 0.1$, (b) $Q_2 = 0.17$, (c) $Q_2 = 0.36$. The first lasing mode intensity is assumed to be $s_1 = (4/3)(w-1)$, which is easily derived from simple calculations. In the case (b), quadratic-to-quartic transition occurs at $w = w_c \simeq 2$.

4.4. BREAKUP OF CW MULTIMODE OSCILLATIONS

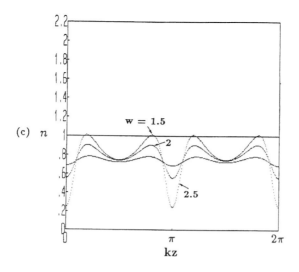

Figure 4.26: (continued).

4.4.4 Antiphase selfpulsations in the Λ-scheme

When the pump is increased beyond w_c, a peculiar antiphase pulsation appears in the multimode domain, before the single mode domain is reached. The threshold pump for the onset of this instability depends on the pump condition (spot size).

In the strong pump regime and above w_c, in which simultaneous multitransition oscillations are occurring, each modal output exhibits a complex noise power spectrum featuring the enhanced low frequency relaxation oscillation peak such as f_2, f_3 and new spectral peaks at combination frequencies of the relaxation oscillations f_k. However, even in such strong pump regimes, the total output displays an ultimate antiphase dynamic nature as it is free from noise except at the McCumber relaxation oscillation [D-4]. Typical examples of modal (1055 nm mode) and total output waveforms and corresponding power spectra in the two-mode (at 1055 nm and 1060 nm) region, i.e., Λ-scheme, in which two transitions originating from the same upper level and two different lower levels, are shown in Fig. 4.27. The fluctuation amplitude of selfpulsations at f_2 is 10% of the averaged output power.

Frequencies of the sharp peaks appearing in Fig. 4.27 and which are different from f_k are found to have unexpected regularities. Varying the pump power and/or the laser cavity length by shifting the crystal perpendicular to the pump axis produced qualitatively the same spectra.

In the case of three-mode oscillation involving the three transitions,

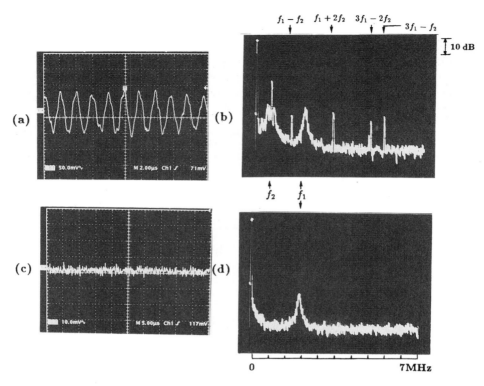

Figure 4.27: Antiphase selfpulsations in the two-mode LD-pumped LNP laser above the Q2Q transition. (a) modal output waveform (2 μs/div.), (c) total output waveform (5 μs/div.), (b) modal output power spectrum, (d) total output power spectrum. The modal output is shown for the 1055 nm mode. The input signal to the rf spectrum analyzer for the total output intensity was attenuated by −10 dB.

the similar selfpulsation at f_2 was observed.

It is well-known that class-B lasers, in which polarization dynamics can be adiabatically eliminated, instabilities never takes place without additional degree of freedom, such as modulation, light injection, saturable absorber etc. Therefore, we have to consider a new class of model to explain observed antiphase selfpulsations at the lower relaxation oscillation frequency. The cruicial points of the present microchip laser include (1) the creation of the coherence of the two lower level atoms by the THz beat wave within the cavity, (2) the scattering by the low spatial frequency (i.e., beat) population grating, and (3) the extremely short photon lifetime of the microchip cavity. Most recently, we derived rate equations including these effects and the strong localization of population inversion at the pumped surface and obtained the Hopf bifurcation analytically. We showed that (1)

4.4. BREAKUP OF CW MULTIMODE OSCILLATIONS

results in selfpulsations and (2) is the origin of antiphase oscillations [A-4]. The simulated selfpulsation and corresponding power spectrum for the two-mode Λ-scheme are shown in Fig. 4.28. The results indicate the selfpulsation at f_2, however, spectral peaks at combination frequencies above f_1 were not reproduced by this model. The more microscopic interaction among dense amplifying atoms is considered to be incorporated to explain the whole dynamics.

In summary, the role of atom-atom interactions is described in the multimode lasing regime of a free-running LNP laser driven by high-density laser-diode pumping. The antiphase selfpulsations are shown to occur in the high-density pumping regime and the theoretical model of atomic coherence in the Λ scheme is presented to explain selfpulsations. The interaction between excited atomic states in lasers treated in this section would pro-

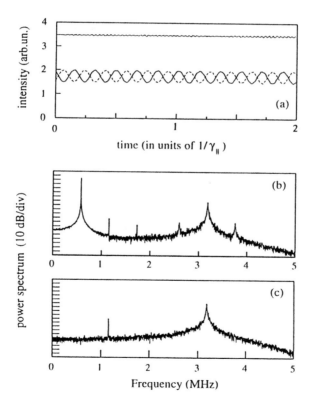

Figure 4.28: Simulated antiphase selfpulsation and corresponding power spectrum. τ (fluorescence lifetime) = 100 μs, τ_p (photon lifetime) = 100 ps. The lifetime of lower levels = 3.2×10^{-10} s. The white noise with the amplitude of 0.002 is added for simulations.

vide a new aspect of laser physics in which the solid-state nature of the active medium begins to play a role. Similar effects may be expected in semiconductor lasers with dense excited carriers.

Another feature of high-density pumping problem is the Auger effect which leads to a high-density correction proportional to N^2 in the evolution of the population inversion. In view of the modal expansion which leads to the usual TSD equations, the N^2 term will induce a second source of recurrence: each modal amplitude $D(n;q,t) = (1/\ell) \int_0^\ell (D^n)z,t) \cos(2\pi qz/\ell)dz$ will depend on $D(n;q \pm q',t)$ as in the TSD equations but also on $D(n+1;q,t)$. Hence a new truncation scheme is necessary to account for the Auger effect. Work along this direction would be an interesting subject for future study.

Keynote Papers

[A]: Section 4.1
1. K. Otsuka and J.-L. Chern, "Variation of Lyapunov exponents on a strange attractorfor spiking laser oscillation", *International Journal of Bifurcation and Chaos*, Vol. 4 (1994) pp. 1111–1114.
[B]: Section 4.2
1. K. Otsuka, J.-L. Chern, and J.-S. Lih, "Experimental suppression of chaos in a modulated multimode laser", *Opt. Lett.*, Vol. 22 (1997) pp. 292–294.
[C]: Section 4.3
1. K. Otsuka, "Laser fluctuation measurements in a bath of strong Gaussian noise", *Appl. Phys.*, Vol. 18 (1979) pp. 415–419.
2. K. Otsuka, "Effects of external perturbations on $LiNdP_4O_{12}$ lasers", *IEEE J. Quantum Electron.*, Vol. QE-15 (1979) pp. 655–663.
3. K. Otsuka, "Nonlinear dynamics in a microchip multimode laser", *SPIE Proc. on Chaos in Optics*, Vol. 2039 (1993) pp. 182–197.
[D]: Section 4.4
1. K. Otsuka, P. Mandel, and E. A. Viktorov, "Breakup of multimode oscillations and low-frequency instability in a microchip solid-state laser by high density pumping", *Phys. Rev. A*, Vol. 54 (1997) [in press].
2. K. Otsuka, R. Kawai, Y. Asakawa, P. Mandel, and E. A. Viktorov, "Simultaneous single-frequency oscillations on different transitions in a laser-diode-pumped $LiNdP_4O_{12}$ Laser", *Opt. Lett.*, Vol. 23 (1998) pp. 201–203.
3. K. Otsuka and T. Yamada, "Resonant absorption in $LiNdP_4O_{12}$ lasers", *Opt. Commun.*, Vol. 17 (1976) pp. 24–27.
4. K. Otsuka, E. A. Viktorov, and P. Mandel, "Atomic interference in a microchip laser operating on a Λ transition", *Europhys. Lett.*, Vol. 45 (1999) pp. 307–313.

References

1) F. Ishiyama and K. Otsuka, International Journal of Bifurcation and Chaos **4** (1994) 1111.

2) H. D. I. Abarbanel, R. Brown, and M. B. Kennel, J. Nonlinear Science **1** (1991) 165.

3) K. Tanaka and Y. Aizawa, Prog. Theor. Phys. **90** (1993) 547.

4) E. Ott, C. Grebogi, and J. A. Yorke, Phys. Rev. Lett. **64** (1990) 1196.

REFERENCES

5) R. Roy, Z. Gills, and K. Scott Thornburg, Optics and Photonics News **5** (1994) 8 and references therein.

6) S. Bielawski, D. Derozier, and P. Glorieux, SPIE Proc. on Chaos in Optics **2039** (1993) 239.

7) T. Tsukamoto, M. Tachikawa, T. Sugawara, and T. Shimizu, Phys. Rev. A **52** (1995) 1561.

8) R. Meucci, W. Gadomski, M. Ciofini, and F. T. Arecchi, Phys. Rev. E **49**(1994) 2528.

9) P. Bryant and K. Wiesenfeld, Phys. Rev. A **33** (1986) 2525.

10) Y. Braiman and I. Goldhirsch, Phys. Rev. Lett. **66** (1991) 2545.

11) R. Lima and M. Pettini, Phys. Rev. A **41** (1990) 726.

12) L. Fronzoni, M. Giocondo, and M. Pettini, Phys. Rev. A **43** (1991) 6483.

13) S. T. Vohra, L. Fabiny, and F. Bucholtz, Phys. Rev. Lett. **75** (1995) 65 and references therein.

14) K. Wiesenfeld and B. McNamara, Phys. Rev. Lett. **55** (1985) 13.

15) F. T. Arecchi, R. Meucci, G. Puccioni, and J. Tredicce, Phys. Rev. Lett. **49** (1982) 1217.

16) R. Lang and K. Kobayashi, IEEE J. Quantum Electron. **QE-16** (1980) 347.

17) K. Ikeda and M. Mizuno, Phys. Rev. Lett. **53** (1984) 1340.

18) M. Mizuno and K. Ikeda, Physica D **36** (1989) 327.

19) K. Ikeda, in *Optical Instabilities*, edited by R. W. Boyd, M. G. Raymer and L. M. Narducci (Cambridge University Press, Cambridge, 1986) pp. 85–98.

20) S. Aubry, in *Solitons and Condensed Matter Physics*, edited by A. R. Bishop and T. Schneider (Springer, Berlin, 1979).

21) K. Otsuka and H. Kawaguchi, Phys. Rev. A **30** (1984) 1575.

22) T. Itoh, S. Machida, K. Nawata, and T. Ikegami, IEEE J. Quantum Electron. **QE-13** (1977) 574.

23) A. E. Siegman, *Lasers*, University Science Books, Mill Valley, CA (1986) pp. 465–466.

24) C. L. Tang, H. Statz, and G. deMars, J. Appl. Phys. **34** (1963) 2289.

25) T. Kimura, K. Otsuka, and M. Saruwatari, IEEE J. Quantum Electron. **QE-6** (1971) 403.

26) M. Blätte, H. G. Danielmeyer, and R. Ulrich, Appl. Phys. **1** (1980) 275.

27) K. Otsuka, T. Yamada, M. Saruwatari, and T. Kimura, IEEE J. Quantum Electron. **QE-11** (1975) 330.

28) H. Koizumi, Acta Crysta. **B32** (1976) 266.

29) K. Otsuka, IEEE J. Quantum Electron. **QE-14** (1978) 1007.

Chapter 5

LASER ARRAY DYNAMICS

Spatiotemporal behavior of spatially distributed systems far from thermal equilibrium is a crucial problem in nonlinear dynamics. In general, high-dimensional systems, different solutions (spatial structures) coexist in stationary state. If one considers systems in which simplified elements are coupled without chaotic response, nontrivial dynamics can never be expected. This is just an initial value problem that results in one of the eigenstates. The Hopfield model is a good example of such systems where different solutions (memory) are associatively recalled depending on initial conditions [1]. On the other hand, in chaotic systems which possesses many equilibria (attractors) as in various dissipative systems, complex dynamical behavior connecting coexisting attractors is expected.

In previous Chapters **2**, **3**, **4**, we discussed nonlinear dynamics in multimode lasers with coupled degrees of freedom. In this chapter, nonlinear dynamics and chaos in laser arrays are studied as a propotypical example of *spatially distributed* nonlinear systems. In section **5.1**, nonlinear dynamics in the *nearest-neighbor coupled* class-A laser array model, that is an optical analogue of discrete complex Ginzburg-Landau systems, are described focusing the transition from homoclinic to heteroclinic chaos [A]. In **5.2**, a realistic model of class-B laser arrays such as semiconductor laser arrays is introduced and their generic dynamics including spot dancing and chaotic itinerancy are described [B]. Finally, nonlinear dynamics in *globally coupled* laser arrays such as antiphase periodic states and instabilities induced by delayed incoherent feedback are described in section **5.3** [C].

5.1 Class-A Laser Array Dynamics

The Ginzburg-Landau systems have been extensively investigated to understand turbulent phenomena in spatially extended nonequilibrium systems. However, attention has been focused on the behavior around the localized equilibrium; the interplay between coexisting equilibria has been left an open question [2].

This section investigates spatiotemporal dynamics in a spatially coupled nonlinear oscillator system, e.g., the discrete complex time-dependent Ginzburg-Landau (CTDGL) equation. As a nonlinear oscillator, let us consider class-A lasers, in which both polarization and population inversion dynamics can be adiabatically eliminated.

5.1.1 Optical analogue of discrete time-dependent complex Ginzburg-Landau equations

Consider coupled class-A laser systems with complex coefficients, assuming $\hat{E}_i = \tilde{E}_i \exp(j\omega t)$, where ω represents a lasing frequency of each emitter in the absence of coupling. Then, dynamics are governed by the following equation:

$$\dot{\tilde{E}}_i = (1/2)[(P - P_{th})/|\tilde{E}_i|^2 - 1/\tau_p](1 - j\beta)\tilde{E}_i \\ + (\alpha_1 + j\alpha_2)[\tilde{E}_{i+1} + \tilde{E}_{i-1} - 2\tilde{E}_i]. \tag{5.1}$$

Here, P is the pump power (P_{th}: threshold), τ_p the photon lifetime, β the on-site nonlinearity, and α_1 and α_2 the coupling coefficients. On-site dispersive nonlinearity appears when the laser is detuned from the gain spectrum peak, but in general this can be introduced into the equation by preparing dispersive elements within the cavity independently. We also assume complex coupling, i.e., phase as well as amplitude coupling, between emitters.

By mathematical manipulation, Equation (5.1) can be generalized into the following well-known *discrete* complex time-dependent Ginzburg-Landau equations (CTDGL) (5.2) without changing the essential physics. Indeed, the following phenomena do occur in the original coupled laser array equation Eq. (5.1) in weakly-coupled regimes. Therefore, we investigate nonlinear dynamics in such an optical analogue of CTDGL systems governed by the equation,

$$\dot{\tilde{E}}_i = \mu_i \tilde{E}_i + (\alpha_1 + j\alpha_2)[\tilde{E}_{i+1} + \tilde{E}_{i-1} - 2\tilde{E}_i] \\ - (1 - j\beta)|\tilde{E}_i|^2 \tilde{E}_i, \quad i = 1, 2, \ldots, N, \tag{5.2}$$

in which we employ the periodic boundary condition, i.e., $\tilde{E}_{N+1} = \tilde{E}_1$. If one sets $\mu_j = 0$ and replaces the last term in Eq. (5.2) by $(1/2)[(P - P_{th}/|\tilde{E}_i|^2 - 1/\tau_p](1-j\beta)\tilde{E}_i$, the orignal coupled class-A laser array equation (5.1) is derived without loss of generality.

The CTDGL equation has been investigated to gain understanding of turbulent phenomena in spatially coupled nonlinear oscillator systems [2–3]. However, attention has been focused on local chaos around the in-phase locking state (i.e., relative phases $= 0$) in an infinite number of elements or even in a continuum limit [3]. Let us investigate CTDGL systems from a small system size limit (i.e., $N = 3$), paying special attention to dynamics involving multiple attractors, and show how dynamics change as the system size increases and the equation approaches the continuum limit.

First, we look for the equilibria of Eq. (5.2) with $\dot{E}_i = 0$ and $\phi_{i+1} - \phi_i \equiv \Delta\phi_i = constant$ independent of i. With these locking conditions, the steady-state solutions given by $\bar{E}_{i+1}^\ell = \bar{E}_i^\ell = \sqrt{\mu + 2\alpha_1[\cos(\Delta\bar{\phi}_i^\ell) - 1]}$, $\Delta\bar{\phi}_i^\ell \equiv (\phi_{i+1} - \phi_i)^\ell = 2\pi\ell/N$ and $\Omega^\ell = \omega - (2\alpha_2 - \beta\bar{E}_i^2) + 2\alpha_2\cos(\Delta\bar{\phi}_i^\ell)$, in which $\ell = 0, 1, \ldots, N-1$, are deduced as equilibria. The linear stability analysis reveals that the equilibria for $N \gg 1$ are not always stable in weakly coupled regimes, i.e., $\beta \gg \alpha_1, \alpha_2$, and instability occurs near the frequency of the energy transfer between elements, α (i.e., α_1, α_2).

5.1.2 $N = 3$ low dimensional chaos: Complex Ginzburg-Landau attractor

A. Symmetry-breaking and symmetry-recovering crisis

Let us consider the simplest case of $N = 3$, which will provide the base for larger systems. For $N = 3$, we have three solutions with $\ell = 0, 1, 2$. Hereafter, we call these solutions the σ^0, σ^+ and σ^- states. These three steady-states coexist only when $\alpha_1 < \mu/2[1 - \cos(2\pi/3)] \equiv \alpha_c$ is satisfied. (For example, $\alpha_c = 1/3$ if $\mu = 1$.) Above this critical coupling α_c, the $\sigma^\pm$ states cannot exist as equilibria because $\bar{E}_i^2$ becomes negative. Therefore, we restrict the following discussions to the regime below α_c to investigate dynamics involving three different equilibria. These two states $\sigma^\pm$ produce the same intensity outputs; however, homodyne or heterodyne measurements can, in principle, distinguish them in experiments.

When the on-site nonlinearity β is increased from zero, a symmetric steady-state σ^0 that is stable in time is realized at first, while the $\sigma^\pm$ states are dynamically unstable. When β increases above the first threshold $\beta_{th,1}$, the σ^0 state (node) becomes unstable and the σ^s states, where one of the relative phases equals zero, appear as unstable fixed points (saddles). These saddle points are also equilibria of Eq. (5.2). Unlike the $\sigma^{0,\pm}$-fixed points, other relative phases $\pm\Delta\phi_c$ depend on the μ, $\alpha_{1,2}$ and β values. These states are easily destroyed by an extremely weak perturbation and make a transition to a stable asymmetric steady state (focus), i.e., σ^+ or σ^-, featuring relaxation oscillations, depending on initial conditions (symmetry-breaking). As β increases further, the asymmetric steady state becomes dynamically unstable and leads to asymmetric chaos via period doubling bifurcation. Figure 5.1(a) shows a bifurcation diagram for $\Delta\phi_2$, assuming $\mu_i(i = 1, 2, 3) \equiv \mu = 1$, and $\alpha_1 = \alpha_2 \equiv \alpha = 0.1$. (Other relative phases

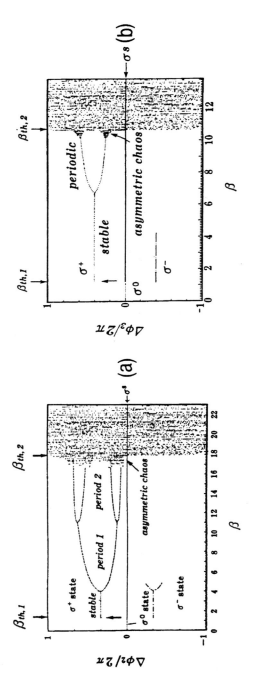

Figure 5.1: Bifurcation diagram for a three-element complex TDGL system. Above $\beta_{th,1}$, only σ^+-state is depicted. $\mu = 1$, $\alpha = 0.1$.

5.1. CLASS-A LASER ARRAY DYNAMICS

have the same bifurcation diagram.) For a given on-site nonlinearity β, an initial condition is chosen and successive maxima and minima of $\Delta\phi_i(t)$ are plotted for $150 < t < 200$ after omitting transients.

As β increases up to $\beta_{th,2}$, where one of the relative phases crosses the $\Delta\phi_i = 0$ line, forming the σ^s saddle point, the chaotic band suddenly widens. Such sudden changes in chaotic attractors, i.e., so-called *crises*, resulting from the collision between a chaotic attractor and an unstable fixed point or periodic orbits, are predicted by Grebogi, Ott and Yorke in various chaotic systems [4–6].

What is happening after the symmetry-recovering crisis in the present system? To invesigate this problem, the temporal evolution of relative phases is calculated. A typical example is shown in Fig. 5.2, which shows waveforms before and after the crisis. Before the crisis (Fig. 5.2(a)), the relative phase wanders chaotically around the σ^+-fixed point. It should be noted that envelopes exhibit a repetition of successive growth and rapid-decay (see regions Q in Fig. 5.2(a)). In addition to this motion, another motion connecting different saddles σ^s is seen to coexist (see regions R in Fig. 5.2(a)), which will be discussed later. The competition between two motions (internal itinerancy) does not persist and the system makes a sudden transition to a period-two cycle pulsation as shown in Fig. 5.2(a). Such coexistence of a strange attractor and an attracting periodic orbit is investigated by Grebogi, Ott and Yorke in terms of interior crises [5] and the present instability is interpreted in terms of a chaotic transient [6].

After the symmetry recovering crisis, self-induced switching among two attractors occurs. The result is given in Fig. 5.2(b), which shows that the system symmetry is recovered and that the system tends to wander chaotically, covering two asymmetric local attractors via the neighborhoods of saddle points.

When the coupling α increases, the threshold for crisis $\beta_{th,2}$ rapidly decreases, such that $\beta_{th,2} = 7.9$ for $\alpha = 0.15$ and $\beta_{th,2} = 2.7$ for $\alpha = 0.2$, and approaches the first threshold $\beta_{th,1}$. It should be noted that the present heteroclinic chaos ceases when the system is attracted by the period-two cycle orbit belonging to σ^+ or σ^- similarly to Fig. 5.2(a). However, the dwell time within one attractor decreases as β increases and the system tends to be hardly attracted by the period-two cycle orbit. Roughly speaking, a persistent heteroclinic chaos takes place for $\beta \geq 20$ in this case. The interplay between these two motions can also be controlled by changing α_1 and α_2. In general, the Q motion becomes dominant when $\alpha_2 > \alpha_1$, whereas the Q motion is suppressed in the opposite case. In both cases, a persistent asymmetric (homoclinic) chaos as well as a persistent heteroclinic chaos featuring the Q or R motion takes place.

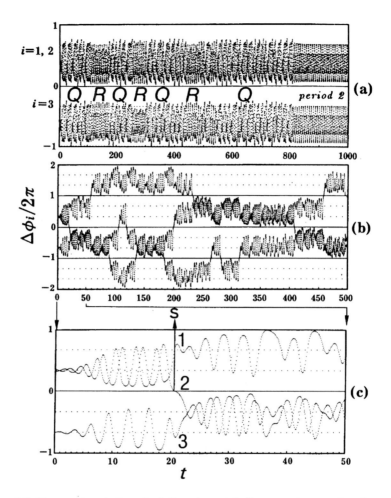

Figure 5.2: Temporal evolution of relative phase $\Delta\phi_i/2\pi$, assuming $\mu = 1$ and $\alpha = 0.1$. (a) Before the symmetry-recovering crisis, $\beta = 17.5$, (b) After the crisis, $\beta = 19.0$. (c) Enlargement of (b).

B. Phase space portrait: Transition from homoclinic to heteroclinic chaos

To understand the above switching dynamics more clearly, the phase trajectory is examined. The phase space trajectories projected onto $[\Delta\phi_2, \Delta\phi_1]$ corresponding to Figs. 5.2(a) and (b) before and after the symmetry-recovering crisis are depicted in Figs. 5.3(b) and (c), where the result for stable regimes after symmetry-breaking is also shown in Fig. 5.3(a) to identify the configuration of fixed points. The points W cor-

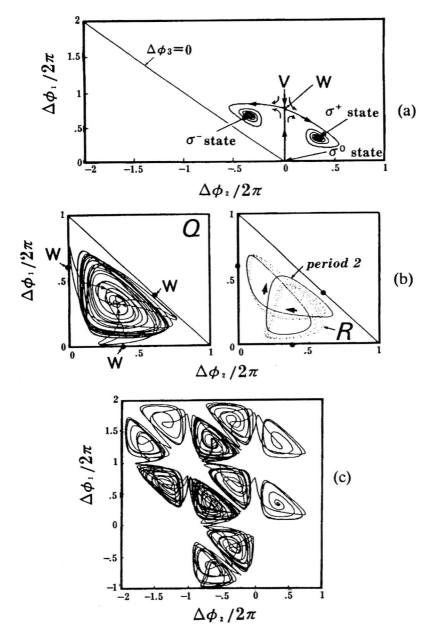

Figure 5.3: (a) Trajectory for stable regimes of $\beta = 2$. $N = 3$, $\mu = 1$, $\alpha = 0.1$. (b) Trajectory corresponding to asymmetric (homoclinic) chaos of Fig. 5.2(a) before crisis. (c) Trajectory corresponding to heteroclinic chaos of Fig. 5.2(b) after crisis.

respond to the previously mentioned σ^s saddle points, and each manifold V of $\sin \Delta\phi_i = 0$ forms the boundary separating the basins of attraction of two attractors. The focus character of $\sigma^\pm$ fixed points is clearly seen in stable regimes (Fig. 5.3(a)). In homoclinic chaos regimes like Fig. 5.2(a), these fixed points become unstable and the trajectory surrounded by the three saddles W consists of a diverging flow (expanding spiral) and a reinjection process into the neighborhood of the σ^+ or σ^- fixed point (see left), as well as another motion connecting different saddles (see right). The trajectories in Fig. 5.3(b) are calculated in region Q ($0 < t < 100$) and region R ($120 < t < 170$) of Fig. 5.2(a), where the period-two cycle orbit is also shown. The former motion implies that $\sigma^\pm$ fixed points possess a saddle-focus nature in the phase space. The successive growth and rapid-decay of waveform-envelopes in the Q regions of Fig. 5.2(a) corresponds to this inherent reinjection process. The inter-saddle motion corresponds to the waveforms in the R regions of Fig. 5.2(a).

This reinjection process stemming from the saddle-focus character strongly suggests the existence of a Shil'nikov-type orbit [7] which has been observed in Belousov-Zhabotinskii reactions [8] and in a laser with feedback [9]. It should be noted that the reinjection process in the present system occurs via three coexisting saddles σ^s, as is seen in Fig. 5.3(b). In heteroclinic chaos regimes (Fig. 5.3(c)), one relative phase crosses over V lines erratically. This heteroclinic symmetric chaos after the crisis, namely *complex Ginzburg-Landau chaos*, differs from the well-known Lorenz (heteroclinic) chaos *since the reinjection process within one asymmetric attractor* seen in Fig. 5.3(c) is absent in the Lorenz chaos as discussed in Chapter **1**.

5.1.3 Self-induced path formation among local attractors

A. Bifurcation scenario

Next, let us consider the dynamics when the system size N is increased. The bifurcation scenario is as follows. As the on-site nonlinearity β is increased from zero, the σ^0 state is realized stably in time up to the first threshold $\beta_{th,1}$. In the region just above $\beta_{th,1}$, spatiotemporal chaos (STC) which is not captured by any particular local attractor, reflecting unstable motions around a basin boundary σ^s-state, which will be discussed later, appears for quite a while. Hereafter, we call this instability global STC. As β is increased beyond this region, two states $\ell = (N-1)/2$ (i.e., σ^+) and $\ell = (N+1)/2$ (σ^-) survive whereas other solutions become unstable foci for *odd N* as will be shown later. For *even N*, only the $\ell = N/2$ (σ^π) attractor survives. The present bifurcation scenario generally exists in weakly coupled regimes independently of the system size N.

A bifurcation diagram for $N = 5$ is shown in Fig. 5.1(b), where $\mu = 1$ and $\alpha_1 = \alpha_2 \equiv \alpha = 0.1$ are assumed. Above the first threshold $\beta_{th,1}$, one of the above two states is realized stably, depending on initial conditions (symmetry-breaking) and exhibits Hopf bifurcation leading to asymmetrical local chaos. Unlike the $N = 3$ case, the period-doubling scenario does not exist for $N \geq 5$. As β is increased up to the second threshold, cooperative switching among the two attractors takes place via the σ^s state. When one increases β in the region above $\beta_{th,2}$, dwell times within the local attractors decreases.

As α (and/or N) is increased, $\beta_{th,2}$ rapidly approaches $\beta_{th,1}$ such that $\beta_{th,2} = 16.5$ (for $\alpha = 0.08$) → $\beta_{th,2} = 10.5$ ($\alpha = 0.1$) → $\beta_{th,2} = 5.8$ ($\alpha = 0.12$), in the case of $N = 5$ for instance. At the same time, the above mentioned global STC region spreads.

B. *Unstable motions at basin boundaries*

Now let us consider the σ^s state for $N > 3$. The σ^s state, where one of the relative phases $\Delta\phi_i(t)$ equals zero, dramatically changes its nature and exhibits *synchronized* unstable motions such that $\tilde{E}_1(t) = \tilde{E}_N(t)$, $\tilde{E}_2(t) = \tilde{E}_{N-1}(t)$, $\tilde{E}_3(t) = \tilde{E}_{N-2}(t)$, etc., as N increases. The σ^s state for $N = 5$ is shown in Fig. 5.4(a), where $\Delta\phi_2(t) = 0$, other relative phases satisfy $\Delta\phi_4(t) = -\Delta\phi_5(t)$ and $\Delta\phi_3(t) = -\Delta\phi_1(t)$. When the σ^s state is excited, two of the relative phases, i.e., $\Delta\phi_1(t)$ and $\Delta\phi_3(t)$ rapidly separate symmetrically. This results from the fact that the two phases $\phi_1(t)$ and $\phi_2(t)$ are synchronized and rotate more rapidly than other phases $\phi_3(t) = \phi_5(t)$ and $\phi_4(t)$. This implies that the intensity of two adjacent elements is increased and the phase rotation speed of these elements is increased accordingly as a result of on-site nonlinearity β. In short, pair phase "rotors" are created.

Here, it should be noted that the averaged value of $\Delta\phi_4 < -\pi$ belongs to the σ^+ attractor and that of $\Delta\phi_5 > \pi$ belongs to the σ^- attractor. If one applies an extremely weak symmetry-breaking noise to this "hybrid" state, one of the two attractors can be excited in the asymmetric chaos regimes. Therefore, this state definitely forms a basin boundary between the two chaotic attractors similary to the case of $N = 3$.

When N is increased further, the σ^s state exhibits quasi-periodic (torus) and even chaotic motion. (The chaotic σ^s state appears when $N \geq 17$ for $\alpha = 0.1$.) The σ^s state exists independently of the system size. This "hybrid" nature, where one relative phase equals 0, $(N-3)/2$ of the relative phases belong to the basin of one attractor ($> \pi$), and the rest belong to the other attractor basin ($< -\pi$), remains in the regimes in which the σ^s state does not exhibit well-developed chaos. For $\alpha = 0.1$, such a well-developed chaotic σ^s appears roughly above $N = 19$. The σ^s motion for $N = 19$ is shown in Fig. 5.4(b). When the coupling α is increased, the well-developed chaotic σ^s state appears for a smaller N. Inherent unstable motions near basin boundaries may presumably be created by reflecting

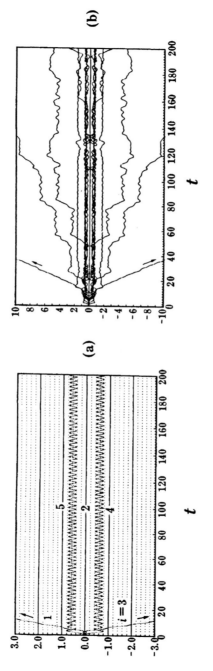

Figure 5.4: Unstable motion of basin boundaries σ^s. $\mu = 1$, $\alpha = 0.1$ and $\beta = 11$. (a) $N = 5$, (b) $N = 19$. These motions do not depend on the on-site nonlinearity β if N and α are fixed.

5.1. CLASS-A LASER ARRAY DYNAMICS

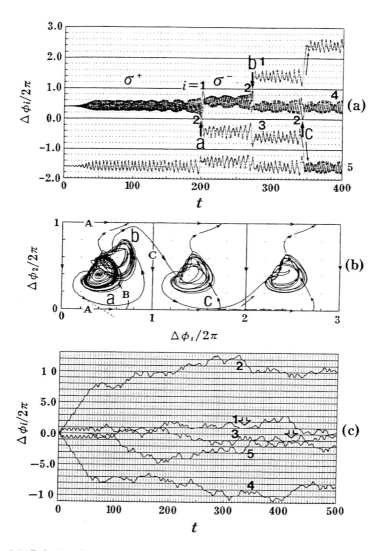

Figure 5.5: Relative-phase dynamics in $N = 5$ CTDGL system. $\mu = 1$. (a) Majority-rule based switching among $\sigma^{\pm}$ attractors. $\alpha = 0.1$, and $\beta = 10.5$. (b) Corresponding phase space trajectory projected on $[\Delta\phi_1, \Delta\phi_2]$. Three σ^s motions **A, B** and **C** projected on this plane are also shown. (c) Global spatiotemporal chaos (STC) around the σ^s state. $\alpha = 0.15$ and $\beta = 2$.

the unstable character of destabilized coexisting equilibria (namely, the *attractor complex*) above the first threshold and provide a possible origin of the previous global STC. Needless to say, there are N equivalent σ^s states.

A similar synchronized chaotic motion also appears for *even N*.

C. Majority rule and memory recovery

An example for $N = 5$ above the second threshold $\beta_{th,2}$ is shown in Fig. 5.5(a). It should be noted that successful cooperative switching among the two attractors is established when the majority rule is satisfied (see points a and b), while switching fails when the condition is not met (see point c). In short, successful switching occurs when the majority of relative phases cross over the $\sin \Delta \phi_i = 0$ line and fall into the basin of attraction of a new attractor just one relative phase crosses over $\pm 2n\pi$. (At point a, for example, just after $\Delta \phi_3$ crosses over zero, $\Delta \phi_{1,4}$ and $\Delta \phi_5$ cross over π and $-2\pi + \pi$, respectively.) This is reasonable in the sense that the system can change the winding number more easily by shifting the rest (i.e., "minority") of the relative phases to the new one rather than by shifting the majority of relative phases back to the initial one once again so as to satisfy the periodic boundary condition. The corresponding phase space trajectory projected onto $[\Delta \phi_1, \Delta \phi_2]$ is shown in Fig. 5.5(b). Switching between the two attractors is clearly seen. Three involving σ^s motions projected on this plane are also shown in the figure. The *majority rule*-based cooperative switching between the two attractors has been found to arise independently of system size if $N(= odd)$ is not so large. This may be due to the fact that "information" spreads over the whole system bidirectionally in a time scale which is shorter than the fluctuation period $\propto 1/\alpha \gg N$.

Let us examine the switching process more precisely. At points a and b, $\Delta \phi_2$ approaches zero and 2π, respectively, and the $i = 1$ and $i = 2$-th elements tend to be locked. As a result, the intensities of the two elements are increased just as in mutual injection locking in lasers and pair "rotors" are created transiently.

When such appropriate pair rotors are created, $\Delta \phi_{1,3}$ separate, the majority condition for switching is satisfied, and switching is established by successful crossing of the basin boundary. To be more specific, as can be seen in Fig. 5.5(b), the system approaches σ^s state **A** in which $\Delta \phi_2 = 0$ at point a, by creating pair rotors. Then, the system is repelled by **A** and crosses successfully over the basin boundary corresponding to σ^s state **B** with $\Delta \phi_3 = 0$. At point b, the system switches back to the σ^+ attractor by crossing over σ^s state **C** with $\Delta \phi_1 = 0$ after being repelled by σ^s-state **A**. In some cases, the system switches to the other attractor by crossing over the basin boundary directly, like the dotted curve shown in the figure.

At point c, $\Delta \phi_{3,1}$ separate when $\Delta \phi_2 \to 0$ (i.e., σ^s state **A**). In this case, however, the pair "rotors" are so strong that these two relative phases again across -2π and 4π successively (see Fig. 5.5(a)), and the system returns to the basin of attraction of σ^+ as shown in Fig. 5.5(b). In some

5.1. CLASS-A LASER ARRAY DYNAMICS

cases, longer-lived pair rotors are created and the system approaches the σ^s state asymptotically, like the dashed curve shown in the figure. At this moment, a similar "hybrid" state featuring the attractor complex motion is excited. In short, in this approach toward the σ^s state (i.e., one relative phase $\to 0$) shown in Fig. 5.4(a), two separating relative phases periodically cross nearly equivalent relative phase values corresponding to the initial attractor, another relative phase remains within the initial attractor, and the rest approaches the value belonging to the other attractor. This implies that the majority of relative phases falls into the basin of the initial attractor almost periodically. This 'history-dependent' motion may return the system to the initial attractor via a path which is periodically opened (see Fig. 5.5(b)) and the initial memory (attractor) is recovered before complete memory blackout, i.e., before the complete "hybrid" state σ^s is established. (At point c of Fig. 5.5(b), the system returns to the initial attractor via a path opened at the *first* cycle, i.e., just after $\Delta\phi_1$ crosses over 4π.) The probability for realizing perfect σ^s-type locking during the course of temporal evolution is extremely low in weakly-coupled regimes. Consequently, this memory-recovering path is always established in the regime where the σ^s state does not exhibit well-developed chaos. Which path (a, b or c) is formed critically depends on how two adjacent element phases approach, i.e., the property of pair rotors.

The symmetrical relative phase separation also occurs due to random creation of single phase rotor during the course of temporal evolution. Figure 5.6 shows temporal evolutions of $\Delta\phi_i$, ϕ_i and E_i^2 for such a case. It is clear from this figure that the phase increases rapidly for one element, whose intensity becomes large (the phase rotor), and then other phases follow it. In such a case, however, the system excites another attractor complex instead. Similarly to the c motion, the system always returns to the initial attractor. In the case of a single rotor, for example, two *adjacent* relative phases separate symmetrically, one relative phase tends to fluctuate around $\pm 2n\pi + \pi$ (n: integer), $(N-3)/2$ of the relative phases remain within the basin of an initial attractor, and the rest approaches the other attractor basin. In this process, the system always returns to the initial attractor via a path periodically opened similarly to the case of pair rotors.

As the system size N or the coupling α is increased, the basin of attraction of the surviving attractors is drastically decreased. Consequently, dwell times within the local attractors decrease and switching fails more often because of frequent rotor creation. Furthermore, the previously mentioned global STC develops when switching fails. In short, in these regimes, two attractors tend to exist just as small 'islands' in a high-dimensional chaotic sea. Finally, the surviving attractors $\sigma^\pm$ are destroyed, thus leading to the persistent global STC. An example of the persistent global STC around the σ^s state is shown in Fig. 5.5(c) when α is increased to 0.15, as-

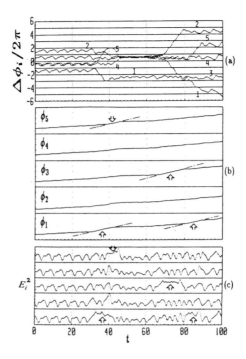

Figure 5.6: Single rotor creation and phase hopping. $N = 5$, $\mu = 1$, $\alpha = 0.1$, $\beta = 13$. (a) relative phase, (b) phase, (c) intensity.

suming $\mu = 1$, $N = 5$ and $\beta = 2$ and setting initial phases near σ^s. During the course of temporal evolution, the σ^s-type locking is also attained for a while as indicated by arrows. The global STC also occurs for *even* N when the σ^π attractor is destroyed.

5.1.4 Intermittent phase turbulence in continuum limit

For double limits of strong coupling and large system size (i.e., $N \gg \alpha \approx 1$), which might correspond to a well-studied continuum CTDGL limit, the dynamics change substantially. The σ^0 state survives, with other attractors being dead, and global STC above $\beta_{th,1}$ tends to be dominated by unstable motions around the σ^0 state as one increases N (α), regardless of whether N is *even* or *odd*. Finally, intermittent phase turbulence around the σ^0 state, featuring discontinuous phase jumping of $\pm 2n\pi$ (n = integer), takes place. This phase jump occurs when E_i of a particular element approaches zero (singular point). At the same time, the phase rotation of this element decreases and a trajectory of $\tilde{E}_i$ approaches the origin at the

5.1. CLASS-A LASER ARRAY DYNAMICS

discontinuous point. A typical example is shown in Fig. 5.7, where $N = 11$, $\mu = 1$, $\alpha = 1$ and $\beta = 3$ are assumed. This is nothing more than the so-called 'hole' solution observed in the standard continuum CTDGL equation [3]. Unlike the weakly-coupled CTDGLs discussed above, such an abrupt phase change occurs only *locally* in space, making it difficult for all the elements to change phase cooperatively. This is reasonable since "information" cannot spread over the whole system within the time scale of the

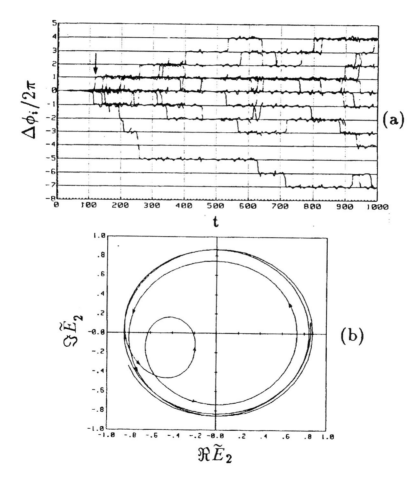

Figure 5.7: Intermittent turbulence in a strongly-coupled regime. $N = 11$, $\mu = 1$, $\alpha = 1$, and $\beta = 3$. (a) relative phase, (b) phase space trajectory of $\tilde{E}_2$ near the discontinuity point indicated by the arrow in (a). The loop structure corresponds to the "hole" creation.

fluctuation $1/\alpha \ll N$ in this case. As β increases, the laminar period decreases and the system tends to exhibit discontinuous phase jumping more frequently.

5.1.5 Switching paths in CTDGLs

On the basis of the computer experiments described so far, the conceptual model of self-induced path formation shown in Fig. 5.8 is obtained for CTDGL systems. In weakly-coupled regimes, a path connecting each attractor (σ^+ or σ^-) and the attractor complex to which it belongs is created for $N(odd) \geq 5$ (see Fig. 5.8(a)), whereas the two homoclinic attractors are simply connected through the saddle for $N = 3$. Switching between the two attractors is governed by the majority rule. Which path is formed critically depends on the character of the phase rotors. Various unstable orbits (attractor complex), which succeed the initial attractor nature, are excited according to the number and property of the phase rotors as well. For *even* N, only the path connecting the σ^π attractor and the attractor complex is formed. (See Fig. 5.8(b).) When one increases N and/or α, the basin of attraction of surviving attractors decreases, thus leading to the global STC featuring attractor complex motions. Finally, in a strongly coupled large system limit (Fig. 5.8(c)), the intermittent phase turbulence

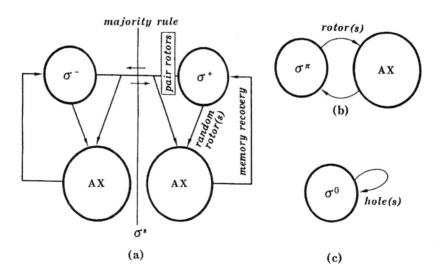

Figure 5.8: Conceptual model of self-induced path-formation in CTDGL systems, where AX denotes the attractor complex. (a) $N = $ odd; weakly-coupled regimes, (b) $N = $ even; weakly-coupled regimes, (c) Strongly-couyled and large system size systems.

around the σ^s state takes place as a result of "hole" creations, regardless of whether N is *even* or *odd*.

A detailed study should be made of fixed points and the dynamics for wider parameter regions and dwell time characteristics within attractors. Additionally, theoretical investigation of the peculiar unstable nature of the σ^s-state and the majority rule dynamics in connection with the system size as well as other parameter values is an interesting future task.

5.1.6 Chaotic itinerancy in open CTDGL systems

In *looped* CTDGL systems with the periodic boundary condition discussed so far, however, local attractors among which switching takes place are restricted to only two symmetric ones with respect to the σ^0 state, although unstable attractor ruins appear around σ^s as N is increased. If the periodic boundary condition, which strongly restricts dynamics and creates the unstable nature near basin boundaries, is removed, more flexible dynamics involving many coexisting patterns are expected. Indeed, a preliminary study of such *open* CTDGL systems indicates self-formation of different patterns. Figure 5.9 shows an example of $\Delta\phi_i(t)$ for an $N = 9$ open CTDGL system in a chaotic itinerancy regime, assuming $\mu = 1$, $\alpha = 0.15$ and $\beta = 8.0$. Unlike looped CTDGL systems, switching takes place by changing relative phases *locally*. To be more specific, $\Delta\phi_{4,5,6}$ change their average values for $a \to b$ switching and $\Delta\phi_{3,4}$ do so for $b \to c$ switch-

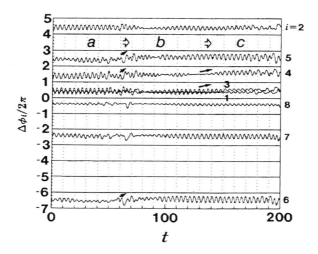

Figure 5.9: Chaotic itinerancy in an open CTDGL system, where the periodic boundary condition is omitted in Eq. (5.1). $N = 9$, $\mu = 1$, $\alpha = 0.15$ and $\beta = 8$.

ing, whereas relative phases for outer elements $\Delta\phi_{1,2,7,8}$ do not change at switching points. Spatial patterns among which chaotic itinerancy occurs are not restricted to these three patterns, although the others are not shown in the figure. It should be noted that some patterns are created during the course of temporal evolution and that they are not destabilized patterns born of the equilibria of the system. This implies that these *new* patterns are born in the strong correlation with the past "history" of the system. A similar new pattern creation has been demonstrated in a coupled-element multistable optical chain model which will be discussed in the next Chapter **6**.

5.1.7 Summary and discussion

High dimensional chaotic dynamics involving multiple attractors are described in this section featuring coupled class-A laser arrays, that is the optical analogue of a discrete complex Ginzburg-Landau equation. A new form of homoclinic (Shil'nikov-type) to heteroclinic chaos transition (complex Ginzburg-Landau attractor) is shown in the three-degrees-of-freedom arrays. Self-induced path formation among local attractors, featuring majority rule dynamics, inherent unstable motions around basin boundaries, and history-dependent memory recovery, is demonstrated for larger-degrees-of-freedom systems. Finally, self-induced pattern formations in open CTDGL systems have been briefly discussed.

Self-induced path formation among local attractors is indeed derived from the chaotic motion generated by the system itself, and the system creates easy paths for switching among attractors utilizing only a few degrees of freedom of the system, i.e., *creative minority*, on the basis of its inherent dynamical rules such as the creation of pair rotors and the majority rule. When switching in a local chaos regime is implemented by externally applied noise without knowing how the system forms a switching path, the probability of finding the switching path may be extremely low. A quantitative evaluation of the effectiveness of 'history-dependent' chaotic force for creating easy switching paths and/or new patterns is an essential common subject to be investigated in high dimensional chaotic systems and should provide new insights into latent powers of chaos.

5.2 Evanescent-Field Coupled Class-B Laser Array Dynamics

5.2.1 Self-induced phase turbulence and chaotic itinerancy

Coupled lasers in general are promising candidates for investigating nonlinear dynamics involving multiple equilibria as described in the previous section. One of the simplest problems is the phase locking problem

5.2. EVANESCENT-FIELD COUPLED CLASS-B LASER ARRAY DYNAMICS

in two coupled laser oscillators. When locking is successful, the two lasers are synchronized and act as one, with a unique frequency and with well-defined phase relationships between the oscillators. When synchronization fails, beat oscillations take place. If many lasers are coupled to each other, much more complex behavior is expected.

In this section, we consider spatially-coupled multiple class-B laser systems as a promising realistic system like laser-diode arrays for investigating our concern in high-dimensional chaotic systems. Ripper and Paoli observed phase coupling (i.e., pure imaginary coupling) between twin-waveguide laser diodes (LD) due to an evanescent field [10]. Otsuka extended their idea to high power *multiple* coupled-waveguide lasers (CWL) [11], and pointed out the interesting analogy between CWL and FM-mode-locked lasers in stationary states [12]. This CWL concept has been demonstrated successfully in laser diode (LD) arrays and their dynamics have been studied very recently [13]. Here, let us see such laser array dynamics and predict some generic dynamics, including a self-induced phase turbulence which results in *emitter partition noise*, chaotic itinerancy between ruins of local attractors (supermodes) resulting from *phase hopping* as well as the super-slow relaxation into equilibria.

5.2.2 Stationary states

In the conceptual model of CWL (Fig. 5.10), each emitter supports a single transverse and longitudinal mode with a frequency ω, when isolated from its neighbors. In the following analysis, we assume class-B laser arrays for which the adiabatic elimination of polarization is valid, but for which the population dynamics should be included. In this case, the temporal evolution of the complex field amplitude $\tilde{E}_i$, assuming $E_i(t) = \tilde{E}_i(t)\exp(j\omega t)$, and of the population inversion N_i in the i-th emitter is described by the coupled equations [14]:

$$\dot{\tilde{E}}_i = \frac{1}{2}[G(N_i) - \frac{1}{\tau_p}](1 - j\alpha)\tilde{E}_i + j\ q[\tilde{E}_{i+1} + \tilde{E}_{i-1} - 2\tilde{E}_i], \qquad (5.3)$$

$$\dot{N}_i = P - \frac{N_i}{\tau} - G(N_i)|\tilde{E}_i|^2, \qquad (5.4)$$

$$\alpha = -(4\pi/\lambda)(\partial n/\partial N)/(\partial G/\partial N), \qquad (5.5)$$

where $G(N_i) = G(N_{th}) + (\partial G/\partial N)(N_i - N_{th})$ is the gain (N_{th} is the threshold population inversion), τ_p is the photon lifetime, τ is the population lifetime, λ is the oscillation wavelength, n is the refractive index, $q = \kappa c/n$ (κ is the coupling coefficient) and P is the pump power. Here α denotes the population dependent refractive index due to anomalous dispersion effect

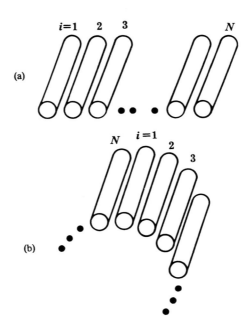

Figure 5.10: Conceptual model of coupled-waveguide lasers (CWL). (a) open CWL, (b) looped CWL.

which is expected in laser diodes, as mentioned in Chapter **1**. This quantity differs from the complex coupling coefficient α_1, α_2 in CTDGL equation (5.2). In usual lasers, α equals zero. Note that pure evanescent-field (i.e., imaginary) coupling is considered and the real coupling α_1 is absent unlike in CTDGL.

First, let's look for equilibria of Eqs. (5.3)–(5.5) with $\dot{E}_i = 0$ and $\phi_{i+1} - \phi_i \equiv \Delta \phi_i = constant$ independent of i, where $\bar{E}_i = E_i \exp(j\phi_i)$. Here, we assume general class-B lasers whose $\alpha = 0$, e.g., solid state, glass and CO_2 lasers. The steady-state solutions are different for open and looped CWL, although they coincide in the limit of $N \to \infty$. In this limit, in-phase as well as π out-of-phase modes with uniform field amplitudes are obtained for both cases.

In the open CWL case (Fig. 5.10(a)), the population inversion distribution becomes nonuniform across the emitters, i.e. $G(N_i) \neq 1/\tau_p$, and the system is found to have solutions which satisfy $\dot{E}_i = 0$, $\dot{N}_i = 0$ with $\bar{\phi}_{i+1} - \bar{\phi}_i \equiv \Delta \bar{\phi}_i = \pm \pi/2$. In short, steady-state values of $\bar{E}_i$ and $\bar{N}_i$ are obtained from the numerical analysis for arbitrary combinations of relative phase $\Delta \bar{\phi}_i$ between adjacent emitters. They are degenerate in lasing frequencies, since $\dot{\phi}_i = 0$ for all the solutions. Examples of these coexisting

5.2. EVANESCENT-FIELD COUPLED CLASS-B LASER ARRAY DYNAMICS

equilibria are shown in Fig. 5.11, where the relative phases are indicated by the rotation angle.

In the case of looped CWL [periodic boundary condition; Fig. 5.10(b)], the population inversion distribution across all the emitters becomes uniform, i.e. $G(N_i) = 1/\tau_p$, because of the coupling between the outermost emitters and the supermodes given by, $\bar{E}_{i+1} = \bar{E}_i$, $\Delta\bar{\phi}_i = 2\pi\ell/N$ and $\Omega^\ell = \omega - 2\alpha + 2\alpha\cos(2\pi\ell/N)$ ($\ell = 0, 1, 2, \ldots, N-1$) are deduced as equilibria with different lasing frequencies Ω^ℓ.

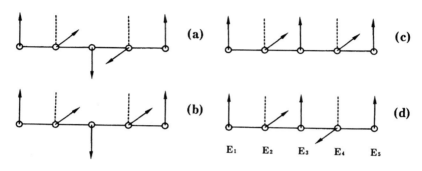

Figure 5.11: Examples of stationary solutions for $N = 5$ open CWL. The phases are expressed by the rotation angles.

5.2.3 Self-induced phase turbulence and spot dancing

Next, let us consider the dynamical properties based on numerical simulations of Eqs. (5.3)–(5.4). Linear stability analysis shows that the equilibra for $N \gg 1$ are not always dynamically stable even if α nonlinearity is absent. Indeed, all solutions for the open CWL are easily destroyed in weakly coupled regimes by an extremely small perturbation to the system. The temporal evolution of relative phase $\Delta\phi_2$ is shown in Fig. 5.12 for the open CWL ($N = 5$), assuming different initial conditions. As can been seen from this figure, there coexist various unstable periodic and chaotic orbits in the phase space. The chaotic evolution of relative phases persists, where the pulsation frequency is approximately given by the energy transfer rate q [14]. At the same time, negligibly small field-amplitude fluctuations approximately given by $(1/N)\sum_{i=1}^{N}(1/2)[G(N_i) - 1/\tau_p]/q$ ($< 10^{-3} \times \bar{E}_{average}$) take place in every emitter, while the total intensity $I = \sum_{i=1}^{N}|\tilde{E}_i|^2$ remains constant. In this sense, the present instabilities can be regarded as self-induced phase turbulence because the intensity fluctuation of individual emitters is negligibly small. This is a sharp contrast to CTDGL in **5.1** where large amplitude fluctuations exist. Such turbulence results in *emitter*

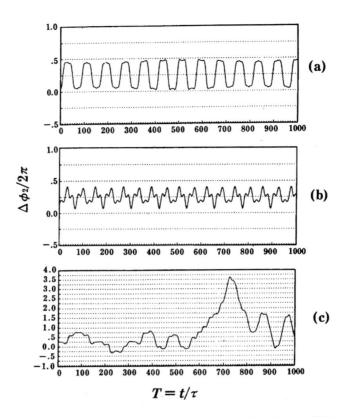

Figure 5.12: Temporal evolution of relative phase $\Delta\phi_2$ for $N = 5$ open CWL, assuming the pump power $p \equiv (1/2)(\partial G/\partial N)N_{th}\tau_p(P/P_{th} - 1) = 0.25$, $q' = q\tau_p = 10^{-4}$ and $\tau/\tau_p = 2 \times 10^3$. Initial conditions are different for three traces. Other relative phases also show periodic or chaotic pulsations.

partition phase noise whose time scale is characterized by energy transfer rate q.

This phase turbulence can be suppressed when population dynamics are omitted, i.e. $\dot{N}_i = 0$. This strongly suggests that the population dynamics play an essential role in this phase turbulence. To clarify this point, the correspondence between phase turbulence and population dynamics is examined. Figure 5.13 shows the temporal evolution of $F \equiv (1/N)\sum_{i=1}^{N}(1/2)[G(N_i) - 1/\tau_p]$, which is the first term of Eq. (5.3), together with that of $\Delta\phi_i$. This quantity F expresses the contribution of population dynamics through Eq. (5.4) and equals zero when phase locking is established. $\bar{F}$ indicated by the arrow in Fig. 5.13(a) corresponds to the steady-state value. The deviation of F from $\bar{F}$ expresses the self-

5.2. EVANESCENT-FIELD COUPLED CLASS-B LASER ARRAY DYNAMICS 223

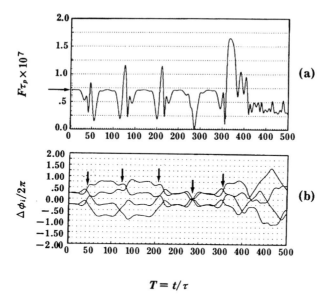

Figure 5.13: (a) Temporal evolution of "destruction force" of phase locking resulting from population inversion dynamics, (b) Corresponding chaotic phase turbulence. Parameter values are the same as in Fig. 5.12(c).

induced "destructive force" of phase-locking, since a stable phase-locking is attained when the first term in Eq. (5.3) remains constant. In short, the population dynamics makes this term fluctuate and acts as a perturbation to phase-locking. Assume a deviation of population inversion from its steady-state value, then the deviation of electric field from the steady-state results, yielding field fluctuation. The electric field fluctuation further increases the deviation of population inversion from the steady-state. This *positive feedback effect* induces the self-induced escape from one equilibrium (local attractor). Indeed, in the case of Fig. 5.13, self-induced switching from one structure (i.e., attractor) to another takes place when the deviation of F from the steady-state peaks. This suggests that self-induced phase turbulence is considered to be a chaotic itinerancy (see Chapter **2**) among destabilized spatial patterns driven by the self-induced destructive force resulting from population dynamics. [In Maxwell-Bloch turbulence described in **2.1**, "chaotic Rabi force" is the origin of chaotic itinerancy.]

If such phase turbulence takes place, the far-field pattern of CWL, which is a Fourier-transform of the near-field pattern, is easily expected to exhibit unstable temporal evolutions. In short, the far-field pattern moves around chaotically in time and deterministic spot dancing develops, even

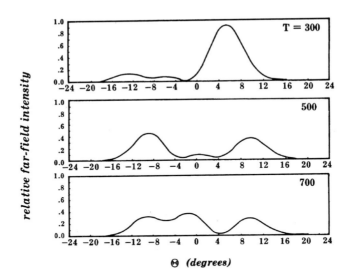

Figure 5.14: Snapshots of far-field pattern corresponding to Fig. 5.12(c), where the half-width of half-maximum of the near-field pattern of each emitter is assumed to be 17 degrees.

though the total near-field intensity is constant. Snapshots of the far-field patterns, corresponding to Fig. 5.12(c) at different times, are shown in Fig. 5.14.

5.2.4 Self-induced chaotic relaxation oscillations and spot dancing in laser-diode arrays

Let us discuss the connection between the present turbulence and instabilities in laser-diode arrays [14] from a practical point of view. If the typical value of nonlinearity, e.g. the α nonlinearity in LDs, is introduced, the π out-of-phase mode (i.e., $\Delta \phi_i \to \pi$), which has been experimentally observed in most LD arrays to date, is found to be realized stably numerically as shown in Fig. 5.15(a). This implies that the adequate amplitude-phase coupling stemming from α nonlinearity suppresses the "intrinsic" phase turbulence shown in Fig. 5.12(c). The favorable relative phase depends on the sign of the α, and the stable in-phase mode is realized when $\alpha < 0$, whereas in real LDs $\alpha > 0$ and the π out-of-phase mode is indeed realized in Fig. 5.15(a). When the coupling coefficient q increases, intrinsic phase turbulence tends to be suppressed. However, in return, instability featuring sustained relaxation oscillations resulting from α nonlinearity becomes dominant and large amplitude fluctuations appear in addition to phase turbulence. Especially, when q (i.e., $\approx$ phase turbulence frequency) is chosen

5.2. EVANESCENT-FIELD COUPLED CLASS-B LASER ARRAY DYNAMICS

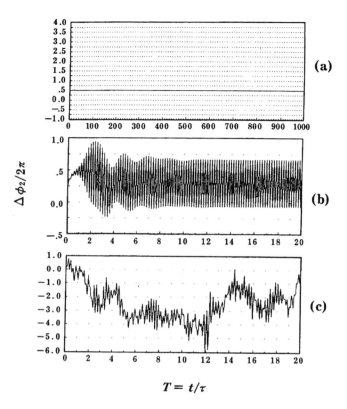

Figure 5.15: The temporal evolution of relative phases in coupled LD arrays for different coupling q. $N = 5$, $\beta = 5$, and other parameters are the same as in Fig. 5.12. (a) $q\tau_p = 10^{-4}$, (b) 2×10^{-4}, (c) 1.78×10^{-3}.

near the intrinsic relaxation oscillation frequency of class B lasers [see Chapter 2], which is given by $f_R = (1/2\pi)\sqrt{(P/P_{th} - 1)/\tau\tau_p}$, periodic and chaotic relaxation oscillations appear, resulting from the amplitude modulation due to the α-induced amplitude-phase coupling. Such amplitude-modulation induced chaotic relaxation oscillations are commonly observed in class B lasers as described in Chapter 2. The simulated relative phase evolutions corresponding to periodic and chaotic relaxation oscillations are shown in Fig. 5.15(b) and (c), respectively, for different coupling q.

If such amplitude as well as phase turbulence take place, the far-filed pattern of laser-diode arrays exhibits spot dancing similarly to Fig. 5.14 in usual class-B laser arrays with $\alpha = 0$. Simulated results for 10-emitter laser-diode arrays indicating periodic and chaotic spot dancing are shown in Fig. 5.16(a). An experimental example of chaotic spot dancing in laser-

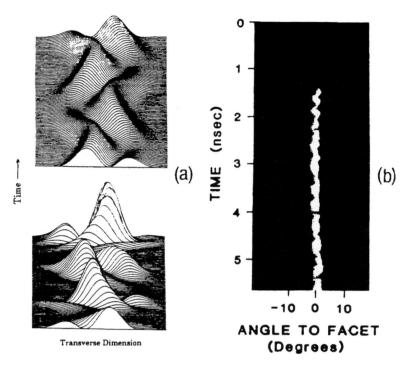

Figure 5.16: (a) Numerical periodic and chaotic spot dancing. (b) experimental chaotic spot dancing observed in an LD array [by courtesy of H. G. Winful, University of Michigan].

diode arrays observed by a streak camera is shown in Fig. 5.16(b).

5.2.5 Chaotic itinerancy and super-slow relaxation

In open CWL depicted in Fig. 5.10(a), eigenmodes are degenerate in lasing energy and persistent phase turbulence occurs featuring chaotic itinerancy among unstable equilibria with $\Delta\phi_i = \pm\pi/2 \pm 2m\pi$ ($m = 0, 1, 2, \ldots$). Conversely in looped CWL with a periodic boundary condition shown in Fig. 5.10(b), the equilibrium bifurcates into S supermodes with different lasing energies where $S = (N + 1)/2$ for N = odd and $S = (N + 2)/2$ for N = even. Consequently, much more complicated behavior is expected to occur resulting from the interplay between nondegenerate supermode solutions. As expected, when the initial conditions are distributed, relative phases wander chaotically between unstable orbits localized around supermodes ("attractor ruins") as is shown in Fig. 5.17(a).

After some transient spatiotemporal chaos (STC), the system is cap-

5.2. EVANESCENT-FIELD COUPLED CLASS-B LASER ARRAY DYNAMICS   227

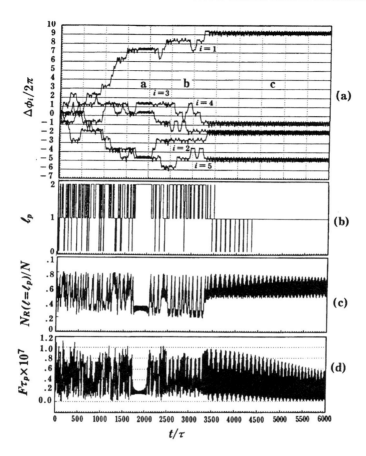

Figure 5.17: Chaotic itinerancy in $N = 5$ looped CWL. Parameter values are the same as in Fig. 5.12. (a) Relative phase, (b) The nearest norm supermode number, (c) The norm from the nearest supermode, (c) The destruction force of phase-locking.

tured by the $\ell = 2$ attractor ruin connected with STC and fluctuates chaotically with a frequency of $\approx q$ in region a. Such localized phase turbulence survives for a long period of time, and then the system jumps out from this local attractor ruin, featuring a transient excitation of two rapidly-rotating phases similar to CTDGL systems, and begins to search for other attractors chaotically while changing relative phases abruptly by an amount of $\approx \pm 2\pi m + 2\pi n/N$ (m, n: $0, 1, 2, \ldots$) in region b. Such an abrupt jump is referred to as *phase hopping* hereafter. (Similarly to CTDGL systems described in section **5.1**, the present system also possesses the synchronized motion like Fig. 5.4(a).) In this regime b, different supermodes are excited in space and seem to compete with each other, resulting in STC. In region c, the system goes by way of the $\ell = 1$ attractor ruin and is finally frozen to the $\ell = 1$ supermode with an extremely long relaxation time ($\approx 10^5 \times \tau$), although this is not depicted in the figure since its damping rate is extremely small. This peculiar slow-relaxation features an initial non-exponential decay followed by an ordinary exponential decay.

It should be noted that the system relaxes to the $\ell = 2$ (and its equivalent $\ell = 3$) supermode solution in an oscillatory fashion if initial conditions are set sufficiently close to steady-state solutions. The $\ell = 0$ supermode also has an extremely narrow basin of attraction with a node character. (This property is independent of system size.) In other words, in the looped CWL case, there are stable fixed points, which have very narrow basins of attraction, and they are surrounded by "ruins" of attractors where local chaotic orbits exist.

To characterize the dynamics in the STC region, we introduce the "norm" from supermode solutions defined by $N_R = \sum_{i=1}^{N} |\cos(\Delta\phi_i) - \cos(2\pi\ell/N)|$ and plot supermode number ℓ_p, whose norm indicates a minimum value, and $N_R(\ell = \ell_p)/N$ in Figs. 5.17(b)–(c), respectively. These figures indicate the connectivity of ruins (i.e., $[\ell = 0] \rightleftharpoons [\ell = 1] \rightleftharpoons [\ell = 2]$) and which supermode the system moves around most closely and the distance from it. There is a clear correspondence between ℓ_p and N_R/N. In short, $\ell_p = 2$ supermode appears at the bottoms of norms, while $\ell_p = 1$ appears at the tops of norms in the STC region. In particular, the system is captured to the $\ell = 2$ ruin only when the norm is adequately small (region a), while it is captured to the $\ell = 1$ ruin even when the norm is relatively large (region c), reflecting the difference in basin of attraction mentioned above. Note that the $\ell_p = 0$ appears only instantaneously, reflecting an extremely narrow basin of attraction, and that chaotic itinerancy occurs dominantly between other attractors, i.e., $\ell = 1$ and 2 as for $N = 5$. In region c, the norm approaches zero very slowly, and finally the $\ell = 1$ supermode is established. Such super-slow damping to equilibria, whose time scale is much longer than other lasing time scales, occurs generally whenever the system can find the basin of attraction. This could be an example of *stagnation phenomena* often observed in high-dimensional chaotic systems [15]. The relaxation time increases when pumping is increased. This result is in sharp contrast to the fact that relaxation time to the steady-state decreases with an increase in pumping in usual population-photon number dynamics (i.e., relaxation oscillations) in which 'phase' dynamics are omitted.

What relationships are there between population dynamics and phase turbulence, especially phase hopping? To investigate this problem, we calculate "destruction force" for phase locking F, similarly to the open CWL. In the looped CWL case, $\bar{F}$ equals zero when phase-locking is established.

5.2. EVANESCENT-FIELD COUPLED CLASS-B LASER ARRAY DYNAMICS

From the temporal evolution of F shown in Fig. 5.17(d), emitter partition noise increases when F increases. Moreover, phase hopping is found to occur at the time when F abruptly increases, although F-peaks are modulated on the time scale of background emitter partition noise. This is the same as the result for open CWL shown in Fig. 5.13.

Extensive numerical simulations show that the probability of the system finding a basin of attraction like region c in Fig. 5.17(a) decreases as N increases, while the number of attractor ruins, among which chaotic itinerancy takes place, increases. Such chaotic itinerancy can be eliminated and phase-locking is established immediately when the coupling coefficient is strong enough. This self-induced phase turbulence is also eliminated if pure real coupling is introduced. The pure real coupling is attained by using a diffraction coupling or a mutual optical injection instead of an evanescent-wave coupling. In such cases, the locking becomes AM-locking, where the equilibria are defined as eigenvalues of the corresponding quantum mechanical harmonic oscillator [16].

It is interesting to compare dynamics occurring in looped CWL with those of CTDGL systems in **5.1**. In CTDGL, the interplay between complex coupling and the on-site nonlinearity is the origin of instability, while in the case of pure-imaginary coupled CWL, population dynamics provides the origin of phase turbulence. This results in the difference in attractors among which switching takes place. As discussed in **5.1**, switching in CTDGL occurs involving attractors having the same oscillation energy, while in CWL it occurs among attractors having different oscillation energies.

The question then arises: *What condition is required to establish successful switching from one attractor to another in the case of CWL?* To answer this question, a numerical result showing successful cooperative switching from the $\ell = 2$ local chaos to $\ell = 1$ local chaos is depicted in Fig. 5.18(a). It should be noted that successful switching is established when the majority of relative phases (i.e., $\Delta\phi_{1,3,5}$ in this case) approach the value of new attractor, i.e. $\Delta\phi_i = \pm 2m\pi + 2\pi/5$ just after one relative phase (i.e., $\Delta\phi_5$) crosses over the $\sin\Delta\phi_i = 0$ line. This is nothing more than the majority rule which governs the switching in CTDGL systems. This implies that the majority rule also comes into existence as for CWL for changing the winding number cooperatively to satisfy the periodic boundary condition similarly to CTDGL. At the same time, "destruction force" F is found to peak at the switching point. The corresponding phase space trajectory projected onto $[\Delta\phi_2, \Delta\phi_5]$ is depicted in Fig. 5.18(b) to indicate "ruins" of two attractors. From this picture, the basin of attraction of the $\ell = 1$ supermode is considered to be larger than that of the $\ell = 2$ supermode as previously expected. It should be noted here that switching failures result from a transient excitation of two rapidly-rotating phases (rotors), when one of the relative phases approaches $\Delta\phi_i = \pm 2n\pi$

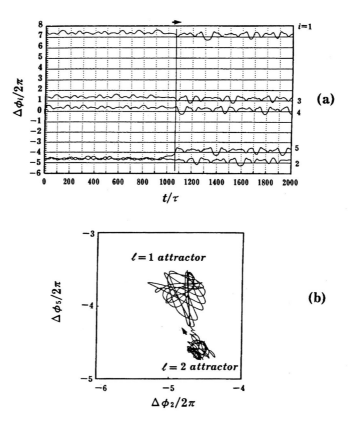

Figure 5.18: An example of cooperative switching from the $\ell = 2$ to $\ell = 1$ chaotic attractors in $N = 5$ looped CWL. Parameters are the same as in Fig. 5.17 and the initial condition is changed. (a) Relative phases, (b) Phase space trajectory.

to change the winding number, similarly to CTDGL systems.

A detailed characterization of complex strange attractors which result in super-slow relaxation is strongly desired.

5.3 Globally-Coupled Class-B Laser Array Dynamics

In this chapter so far, nonlinear dynamics in class-A and B laser arrays, in which each emitter couples each other through *nearest neighbor* coupling. In this section, let us consider a *globally* coupled class-B laser arrays. The conceptual model of proposed globally-coupled laser arrays (GCLA) is illustrated in Fig. 5.19. Laser elements are coupled globally through an external *incoherent* feedback via multiport fiber, polarization

5.3. GLOBALLY-COUPLED CLASS-B LASER ARRAY DYNAMICS

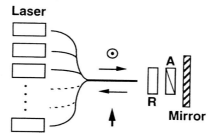

Figure 5.19: A Conceptual model of globally-coupled laser arrays (GCLA) with incoherent feedback.

(Faraday) rotator and external mirror. This GCLA scheme is applicable to linearly polarized lasers.

5.3.1 Factorial dynamical pattern memory in short-delay

The dynamics of GCLA are governed by the following simple rate equations including the cross-gain-saturation by means of reinjected photons. Here, the interefence effect is absent because the polarization direction of feedback field is rotated by 90 degrees with respect to the polarization direction of lasing field.

$$dn_k/dt = w - n_k[1 + s_k + (\gamma/N) \sum_{j \neq k}^{N} s_j(t - t_{k,j})], \qquad (5.6)$$

$$ds_k/dt = K[(n_k - 1)s_k + \epsilon_k n_k + s_{i,k}], \qquad k = 1, 2, \ldots, N. \qquad (5.7)$$

Here, n_k is the normalized population inversion density of the k-th laser element, γ is the coupling strength, $t_{k,j}$ is the time delay, and other nomenclatures are the same as those of multimode laser rate equations (2.24)–(2.26).

First, let us consider the case of short delay, i.e., $t_{k,j} \ll \tau$. In this case, a simple linear stability analysis tells us that the stationary solutions are stable; indicating relaxation oscillations similarly to multimode lasers described in Chapter 2. Almost the same behaviors as multimode lasers in Chapter 2 appear by modulations. For example, if we introduce a deep pump modulation, antiphase periodic states are easily obtained as shown in Fig. 5.20, where $N = 3$ are assumed. Also, dynamical factorial pattern momory is established by means of injection seeding (see Fig. 5.21).

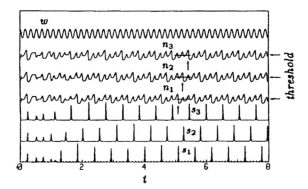

Figure 5.20: Computer analysis for $N = 3$ antiphase periodic states. $w_0 = 2.7$, $\Delta w = 2$, $K = 10^3$, $\tau w_m = 35$ and $\epsilon = 1.2 \times 10^{-7}$.

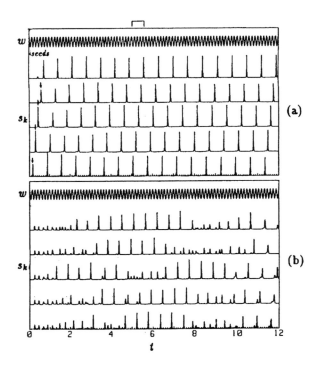

Figure 5.21: (a) Assignment to desired antiphase periodic state pattern by "seeding" in $N = 5$ GCLA. $\tau w_m = 45$ and other parameters are the same as Fig. 5.20. Seed pulse: intensity $s_{i,k} = 0.2$, pulsewidth $\Delta t = 0.06$. (b) Without seeding.

5.3.2 Delay-induced generalized bistability in one-element system

With a delayed feedback, the system represents a system with an infinite temporal dimensions, in which complex dynamics are expected. In this subsection, before treating GCLA let us first consider nonlinear dynamics expected in *one-element* laser with an incoherent delayed feedback (LIDF) as a base for GCLA dynamics. A schematic model is shown in Fig. 5.22(a).

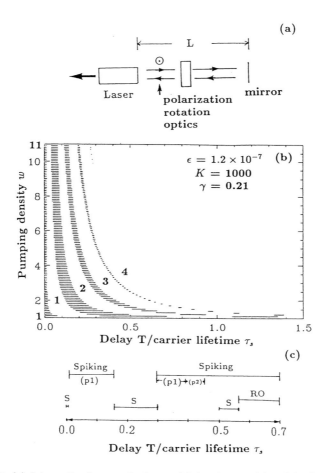

Figure 5.22: (a) Schematic diagam of a laser with incoherent delayed feedback (b) Stability diagram. Stable regions are indicated by hatching and multiple unstable zones are indicated by number.

A. Stability diagram

A linear stability analysis around stationary states, i.e., $\bar{n} \simeq \bar{s}/(\bar{s}+\epsilon)$, $\bar{s} \simeq (w-1)/(1+\gamma) + \epsilon w/(w-1)$ ($\epsilon \ll 1$) easily shows that stationary states become dynamically unstable in some regimes of delay time $t_D \equiv t_{k,j}$. The characteristic equation for infinitesmal deviation, i.e., $n = \bar{n} + \delta n e^{\lambda t}$ and $s = \bar{s} + \delta s e^{\lambda t}$, is given by

$$\lambda^2 + p\lambda + q + r e^{-\lambda t_D} = 0, \tag{5.8}$$

where $p = [1 + (1+\gamma)\bar{s} - K(\bar{n}-1)]$, $q = K[(1+\gamma\bar{s})(1-\bar{n}) + \bar{s} + \bar{n}\epsilon]$ and $r = \bar{n}K(\bar{s}+\epsilon)\gamma$.

The system is stable if all the real parts of λ are negative. The stable regime can be determined by applying the theorem of Bhatt and Hsu [17]. The present linear stability analysis reveals that there are *multiple* stable zones as shown in Fig. 5.22(b).

A detailed analytical investigation of the present system is reported in [18].

B. Toda oscillator model

To provide physical insight into unstable behavior in this system, let us here introduce the Toda potential for laser equations (5.6)–(5.7) for $N = 1$ (single element inocoherent delayed feedback laser). A logarithmic transformation, $u(t) \equiv \ln s(t)$, is introduced. Then, the motion is described by

$$d^2u/dt^2 + \kappa(du/dt) + \partial V/\partial u = F_D, \tag{5.9}$$

where the potential V, the damping constant κ, and the driving force F_D are

$$V = K[w\epsilon e^{-u} + e^u - (w-1)u - (w\epsilon+1)], \tag{5.10}$$

$$\kappa = \epsilon(K + du/dt)/(e^u + \epsilon) + 1 + e^u, \tag{5.11}$$

$$F_D = -\gamma(K + du/dt)e^{u(t-t_D)}. \tag{5.12}$$

In this picture, a dynamic change in photon density can be understood in terms of particle motion in a highly asymmetric laser Toda potential V with a complicated damping constant κ and a retarded driving force F_D. Indeed, in the limit of $\epsilon \to 0$, the original Toda potential is obtained by replacing u by $-u$ [19]. Since $K + du/dt \simeq Kn > 0$ ($\epsilon \ll 1$) from Eq. (5.7) with $N = 1$, we always have $F_D < 0$. This implies that the driving force F_D always serves to apply a force to the system in the negative direction in proportion to feedback light intensity $s(t-t_D)$ as shown in Fig. 5.23(a). On the other hand, the damping constant κ increases as photon density increases. In the original Toda oscillator systems, the damping constant does not depend on u.

Without feedback, i.e., $F_D = 0$, the particle approaches the ground state featuring damped relaxation oscillations. If delayed feedback is introduced, the hamiltonian motion around the ground state (namely, *soft mode*) takes place resulting from the balance of damping force $\kappa(du/dt)$

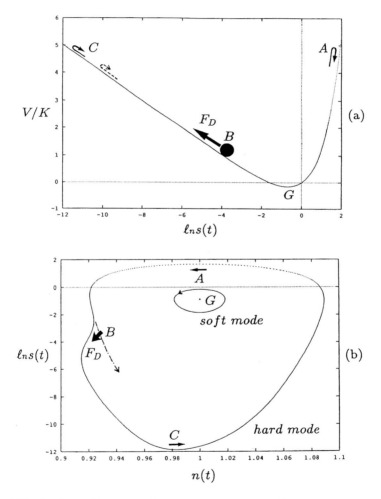

Figure 5.23: (a) Laser-Toda potential. $w = 1.5$, $\gamma = 0.21$, $K = 10^3$, and $\epsilon = 1.2 \times 10^{-7}$. A particle motion in the potential corresponding to the hard mode (spiking mode) oscillation is shown. (b) Phase space trajectories for coexisting hard and soft modes at $w = 1.5$ and $t_D = 0.6$. Other parameters are the same as (a). Without feedback, the spiking mode approaches the fixed point (stationary lasing state) in a spiral fashion as shown by the dotted-dashed curve.

and driving force F_D in unstable regimes predicted by the linear stability analysis described in A, or equivalently by harmonic balance analysis of Eq. (5.9).

In addition to this mode, a spike-like waveform reflecting the asymmetric potential is also expected in large signal regimes. Let us call this the *hard mode* hereafter. This hard mode oscillation corresponds to a regenerative generation of the first spike in the onset of relaxation oscillation which bulids up from a nonlasing solution of Eqs. (5.6)–(5.7) with $N = 1$, i.e., $\bar{n} \simeq w$ and $\bar{s} \simeq 0$ ($\epsilon \ll 1$). This is noting more than the spiking-mode oscillation described in **2.2**, which appears through deep pump modulation of class-B lasers. It should be noted that in the present system the spiking-mode oscillation is excited by a delay-induced 'pulse-like' modulation of

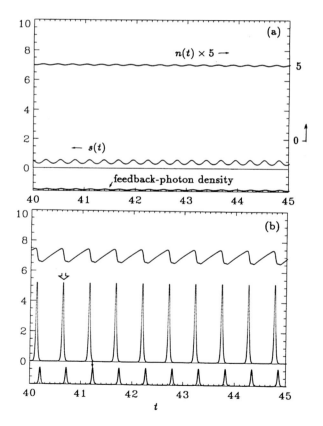

Figure 5.24: Waveforms of the soft (a) and hard (b) mode. Adopted parameters are the same as Fig. 5.23(b).

5.3. GLOBALLY-COUPLED CLASS-B LASER ARRAY DYNAMICS

population inversion, as will be shown later. Which oscillation mode (soft or hard) is excited depends on the initial conditions for appropriate delay times, although regions for getting the spiking mode cannot be determined by small signal linear stability analysis mentioned above. Typical examples of coexisting periodic sustained relaxation osillations (e.g., soft mode) and spiking mode oscillations (e.g., hard mode), are shown in Fig. 5.24. Corresponding phase space trajectories are shown in Fig. 5.23(b).

C. Bifurcation diagram

An typical bifurcation diagram is shown in Fig. 5.25 as a function of the pump w, assuming $K = 10^3$, $\gamma = 0.21$, $\epsilon = 1.2 \times 10^{-7}$ and $t_D = 0.6$. For a given w, succcessive maxima of $s(t)$ are plotted after omitting transients. The coexistence of two different attractors corresponding to the soft and hard modes and generalized bistability are clearly seen in the figure. In bistable regimes, two attractors coexist with different basins of attraction and each motion can be selectively excited by different initial conditions. The soft mode is always attained when initial conditions are chosen near lasing stationary states. The hard mode is exited when the initial conditions are set near non-lasing stationary solutions. Unstable zones for the soft mode determined by the linear stability analysis are shown in the figure.

In the phase space trajectories shown in Fig. 5.23(b), the hard mode

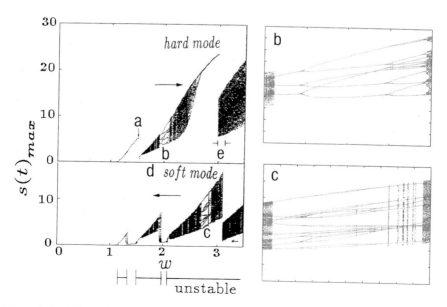

Figure 5.25: Bifurcation diagram for $t_D = 0.6$, assuming the same paramters as Fig. 5.23. Window structures at region **b** and **c** are shown on the right.

exhibits periodic spike-pulse oscillation whose time series includes $s(t) \to 0$ during most of the period. The corresponding particle motion for hard mode in the potential is depicted in Fig. 5.23(a). Due to the spike-wise feedback light, a particle is pushed toward the negative direction by F_D at B and reaches C. Without feedback light, a particle returns by the path shown by the dashed curve in Fig. 5.23(a) and approaches the ground state featuring relaxation oscillation.

When w is increased, each periodic motion exhibits a complicated bifurcation leading to chaos interrupted by periodic orbits as shown in Fig. 5.25. The hard mode features a period-doubling bifurcation similarly to class-B

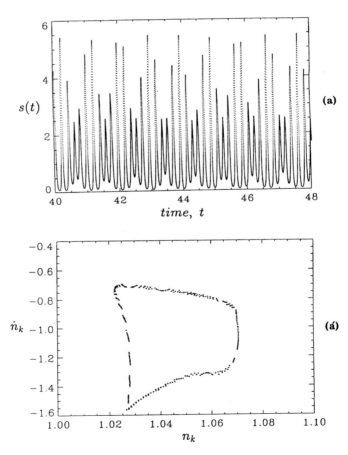

Figure 5.26: (a) Soft mode quasiperiodic motion, (a') corresponding Poincaré section at $w = 2.4$ (b) Hard mode chaotic spiking motion, (b') corresponding Poincaré section at $w = 2.4$.

lasers with deep modulation [20]. In a window region **b**, a clear intermittency, indicative of a saddle-node bifurcation, occurs just below the left boundary, and a period-doubling 3×2^n ($n = 1, 2, 3$) starting from a period-3 spiking mode takes place, as shown in the right upper figure. Then, band chaos with 3×2^n periodicity appears and finally chaos spreads at the right boundary due to crisis.

The soft sustained relaxation oscillation mode, on the other hand, indicates motion on a torus and a quasiperiodic route to chaos featuring commensurate and incommensuate locking as well as windows. An enlargement of window region **c** is shown in the right bottom figure. In this figure, locking begins with period-9, then period-18 appears and returns

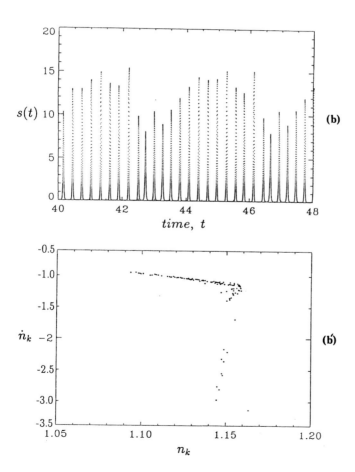

Figure 5.26: (continued).

to period-9. The winding number for this locking remains constant and is determined from Poincaré section to be 7/9.

Temporal evolutions of coexisting chaotic motion born from the hard mode and quasiperiodic motion (soft mode) in bistable regime and corresponding Poincaré sections are shown in Fig. 5.26.

5.3.3 Delay-induced instability in GCLA

A. Linear stability analysis

Let us assume $t_{k,j} = t_D$ for all k and j. The stationary solutions of Eqs. (5.6) and (5.7) are determined by

$$\bar{n}_k \simeq 1, \quad \bar{s}_k = (w-1)/(1+\gamma), \tag{5.13}$$

assuming $\epsilon w \ll 1$ ($\epsilon_k = \epsilon$ for all k). There are other solutions with one or several $\bar{s}_k = 0$, which are unstable. The linear stability analysis is performed by assuming $x = \bar{x} + \delta x$ and $\delta x = \delta \bar{x} e^{\lambda t}$ ($x = n_k$ or s_k) in Eqs. (5.6) and (5.7) with Eq. (5.13). The characteristic equation is given by

$$\lambda^2 + w\lambda + [K(w-1)/(1+\gamma)]e^{i2\ell\pi/N}e^{-\lambda t_D} + K(w-1)/(1+\gamma) = 0, \tag{5.14}$$

where $\ell = 0, 1, 2, \ldots, N-1$. The linear stability is determined by $\ell = 0$ only as that of the $(0,0)$ mode of ref. [21].

With an increase in w, the system shows a self-sustained relaxation oscillation via a Hopf bifurcation at the second threshold $w = w_{th,2}$. The stable regime is determined by satisfying the condition $Re(\lambda) < 0$ for all λ. Results are shown in Fig. 5.27. As is seen from this figure that there exists the third threshold $w_{th,3}$ above which the system becomes stable again. In stable zones, only damped relaxation oscillations are obtained. Unstable motion which aries when $w_{th,2} < w < w_{th,3}$ is numerically confirmed. This stability diagram is universally valid independently of N.

Here, let us consider physical interpretation for delay-induced instability. When the feedback loop is opened and w is increased from zero, the delayed feedback light triggers a relaxation oscillation. The damping oscillation is memorized and then fedback to the laser. Therefore, the population inversion density is intensively modulated by the feedback beam and the laser produces modulated output if the delay is not too long as compared with the damping time constant. This output is again fedback to the return beam further modulates population inversion density. If the growth rate of pulsation by such a positive feedback mechanism balances with the damping rate of pulsation, the laser evolves itself to a state of sustained pulsation at $w = w_{th,2}$. As the pump rate w increases, the relaxation oscillation frequency increases. At the same time, the damping rate increases. As a result, when w approaches $w_{th,3}$, the feedback-induced

5.3. GLOBALLY-COUPLED CLASS-B LASER ARRAY DYNAMICS

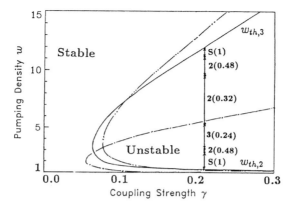

Figure 5.27: Stability diagram of GCLA with incoherent delayed feedback. $\gamma = 0.21$, $\epsilon = 1.2 \times 10^{-7}$. Solid line: $t_D = 0.03$, $K = 1000$, Dotted dashed line: $t_D = 0.05$, $K = 1000$, Two dotted dashed line: $t_D = 0.03$, $K = 800$. S(1) denotes synchronized state, 2 is the 2-clustered state, 3 is the 3-clustered state.

modulation is damped out within the delay time and finally the instability vanishes above $w_{th,3}$.

B. Synchronized pulsation

Let us study dynamics in unstable regimes, which depend on the system size N. The self-sustained pulsation near the boundaries $w_{th,2}$ and w_{th}, which are indicated by S(1), are synchronized and each laser array element exhibits in-phase periodic oscillations. This pulsation features spiking-mode oscillation, e.g., hard mode in the previous subsection. An example of synchronized pulsations and corresponding phase portrait in these domains are shown in Figs. 5.28(a) and (a'). Many random initial values are used to test the synchronization.

Let us provide a physical interpretation for synchronization. Suppose there are initial small inhomogeneous (k-dependent) deviations from stationary values $\bar{n}_k$ and $\bar{s}_k$. From Eq. (5.6), the site dependence of n_k is negligible if the pump rate w is not so large and the feedback photon number $(\gamma/N) \sum_{j=1}^{N} s_j$ is small. Therefore, all n_k are perturbed in the same fashion by this number and all s_k are synchronized through Eq. (5.7).

C. Clustered state

When the pump rate w is increased within unstable zones, feedback intensity $(\gamma/N) \sum_{j=1}^{N} s_j$ increases and the perturbation becomes strong. At this moment, the initial small inhomogeneity in n_k is enlarged by the increased perturbation based on Eq. (5.6). As a consequence, the population dynamics become site dependent and synchronization fails. As soon

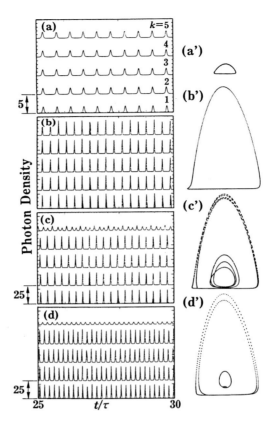

Figure 5.28: Time series (a)–(d) and phase space portraits (a')–(d') for different w assuming $N = 5$, $\gamma = 0.21$, $t_D = 0.03$, $K = 1000$, $\epsilon = 1.2 \times 10^{-7}$. (a)–(a'): $w = 1.3$, (b)–(b'): $w = 2.7$, (c)–(c'): $w = 2.9$, (d)–(d'): $w = 4$.

as synchronization disappears, clustering motion takes place among laser elements as shown in Figs. 5.28 (c)–(c') and (d)–(d'), in which the system is divided into different groups with different synchronized motions similarly to clustered states in multimode laser dynamics [see Chapter 3]. It is interesting to note that the system is divided into hard-mode group and soft-mode group discussed in the previous subsection. It is found that as w is increased, the elements separate one by one from the synchronized state and the system forms clusters consisting of an increased number of groups. In general, there coexist at least $N!/N_1!N_2!\cdots N_k!$ equivalent attractors in the k-clustered state. Because of linear feedback, increasing w implies increasing the depth of modulation. This drives the system to a quasiperiodic or even chaotic clustered state. In the present parameter regime, a quasiperiodic synchronized state is realized just before the onset of clustered state as shown in Figs. 5.28(b) and (b').

The frequency components embedded in the quasiperiodic clustered states provide the key for understanding clustering behavior. The power spectra which correspond to Fig. 5.28 are shown in Fig. 5.29. In the transition process from the synchronized quasiperiodic motion [Fig. 5.28(b)] to the clustered state [Fig. 5.28(c)], the $k = 1, 2, 3, 4$ elements follow the nature of Fig. 5.28(b) with an enhanced quasiperiodicity. To be more specific, the $k = 1, 2, 3, 4$ elements are synchronized while the $k = 5$ element exhibits a different motion (2-clustered state). During this process, the $k = 5$ element is segregated. Note that the power spectra of motions of two groups shown in Figs. 5.29(c) and (c') indicate the locking of $f_2/f_1 = 2/3$ where f_1 is the dominant frequency of the $k = 5$ element and f_2 is the dominant frequency of other synchronized elements. Such a locking among clusters exists in the pump regime up to $w = 3.5$. If the pump rate w is increased further, the modulation depth increases and all clustered elements tend to be locked at the least common frequency components f_3 above the relaxation oscillation frequency as shown in Figs. 5.28(d) and 5.29(d)–(d'),

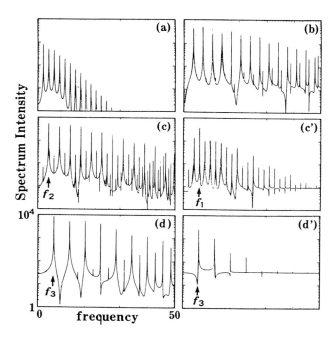

Figure 5.29: Power spectra corresponding to Fig. 5.25 calculated by the fast-Fouier-transform (FFT). (c') and (d'): power spectra for the $k = 5$ element.

where 3-clustered state is realized.

As w is increased further, the inverse process, featuring the unification of all the coexisting attractors, takes place and it leads to a synchronized pulsation just below $w_{th,3}$. Above $w > w_{th,3}$, the system returns to the stable state as shown in Fig. 5.27.

D. Effect of non-equal time delays

Let us consider the effects of multiple different time delays, which are expected in real systems. Because of different delays, feedback beams trigger each element differently. The underlying self-organization mechanism is quite intricate. In Fig. 5.30, realized dynamical states are shown on the plane of the average of randomly chosen time delays versus the standard deviation for $w = 2$ and $N = 7$. Here, hundreds of data are plotted; the square, star, and triangle symbols denote the clustered state, synchronized state and stable state, respectively. The data can be roughly classified into five regimes. In regime A, only the stable state appears. In regime B, the system has the largest probability of achieving synchronized pulsation, which results from an adequate firing on time among the feedback beams in the case of short delays. Synchronization and clustering are mixed with almost equal probabilities in regime C. Regime D is mostly occupied by clustered states. In regime E with a long delay, only the stable state appears again. The existence of regimes A and E indicates the process of attractor fission and fusion described in previous subsection C, and shows again the fundamental characteristic of incoherent feedback.

In the case of arbitrary $t_{k,j}$, waveforms become very complicated. It is

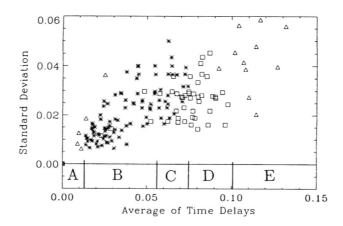

Figure 5.30: Effect of delay time distribution on dynamical states. $N = 7$, $w = 2$, $\gamma = 0.21$, $K = 1000$ and $\epsilon = 1.2 \times 10^{-7}$.

surprising to note that synchronization is replaced by a frequency locking among elements with a fixed pulse phase difference of the order of $\Delta t = 0.01$ in time in the case of short average delay. The process of attractor fission and fusion, which is generic in incoherent feedback, also appears.

Supplement: Gigabit picosecond pulse generation in semiconductor laser with incoherent delayed feedback

If we apply the idea of incoherent delayed feedback scheme to laser diodes, high-speed optical pulse generation is expected since relaxation oscillation frequency in LD's ranges from several GHz to several 10 GHz. Figure S shows examples of delay-induced pulsations for realistic parameters of $\gamma = 0.4$, $\beta \equiv N_0/N_{th} = 0.375$ (N_0: carrier density to reach quasi-Fermi level zero, N_{th}: threshold carrier density), here, delay time scaled by carrier lifetime is 0.1, $K = 1000$, and normalized nonlinear gain $g \equiv G^{(3)}\Gamma V_0/A_g = 5$, where $G^{(3)}$ is the third-order nonlinear gain, Γ is the confinement factor, V_0 is the optical mode volume, and A_g is the differential gain coefficient. Time t is normalized by the carrier lifetime.

If a distributed feedback LD or a surface-emitting LD is combined with a compact polarization rotation optics (e.g., Faraday rotator) and a feedback mirror, dynamic single-mode picosecond optical pulse generations are strongly expected at Gbit/s repetition rates without using any modulation devices [C-4]. Experimental demonstrations are strongly desired.

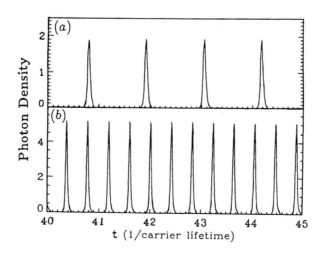

Figure S: Numerical gigabit-rate optical pulse generation in laser diode with incoherent delayed feedback.

Keynote Papers
This Chapter is written on the basis of the following papers:
[A]: Section 5.1
1. K. Otsuka and K. Otsuka, "Global chaos in a discrete time-dependent complex Ginzburg-Landau equation", *OSA Proc. Nonlinear Dynamics in Optical Systems*, N. B. Abraham, E. Garmire, and P. Mandel Eds., Vol. 7 (1990) 570–573.
2. K. Otsuka, "Transition from homoclinic to heteroclinic chaos in coupled laser arrays", *SPIE Proc. on Nonlinear Optics and Materials*, Vol. 1497 (1991) 300–312.
3. K. Otsuka, "Self-induced path formation among local attractors and spatiotemporal chaos in a complex Ginzburg-Landau equation", *Phys. Rev. A*, Vol. 67 (1991) 1393–1396.
[B]: Section 5.2
1. K. Otsuka, "Self-induced phase turbulence and chaotic itinerancy in coupled laser systems", *Phys. Rev. Lett.*, Vol. 65 (1990) 329–332.
2. K. Otsuka, "Complex dynamics in coupled nonlinear element systems", *Int. J. Modern Phys.*, Vol. B5 (1991) 1179–1214.
[C]: Section 5.3
1. J.-L. Chern, K. Otsuka and F. Ishiyama, "Coexistence of two attractors in lasers with incoherent delayed feedback", *Opt. Commun.*, Vol. 96 (1991) 259–266.
2. K. Otsuka and J.-L. Chern, "Dynamical spatial-pattern memory in globally coupled lasers", *Phys. Rev. A*, Vol. 45 (1991) 8288–8291.
3. K. Otsuka and J.-L. Chern, "Synchronization, attractor fission, and attractor fusion in a globally coupled laser array system", *Phys. Rev. A*, Vol. 45 (1991) 5052–5055.
4. K. Otsuka and J.-L. Chern, "High-speed picosecond pulse generation in semiconductor lasers with incoherent delayed feedback", *Opt. Lett.*, Vol. 16 (1990) 1759–1761.

In relation to [C], laser polarization dynamics are reported in
a. K. Otsuka, "Oscillation properties of anisotropic lasers", *IEEE J. Quantum Electron.*, Vol. QE-14 (1978) pp. 49–55.
b. K. Otsuka and H. Iwasaki, "Stabilization of oscillating modes in a $LiNdP_4O_{12}$ laser", *IEEE J. Quantum Electron.*, Vol. QE-12 (1976) pp. 214–217. [This is the first report on dynamics and self-phase-locking type behavior in anisotropic lasers with an intracavity quarter-wave plate, which is the current issue in laser polarization effects in semiconductor lasers; see for example, W. H. Loh, Y. Ozeki, and C. L. Tang, Appl. Phys. Lett. **56** (1990) 2613.]
c. K. Otsuka, K. Kubodera, anf J. Nakano, "Stabilized dual-polarizationn oscillation in a $LiNd_{0.5}La_{0.5}P_4O_{12}$ laser", *IEEE J. Quantum Electron.*, Vol. QE-13 (1977) pp. 398–400.

References

1) J. J. Hopfield, Proc. Natl. Acad. Sci. USA **79** (1982) 2554.

2) See for example, Y. Kuramoto, *Chemical Oscillations, Waves, and Turbulence*, edited by H. Haken (Springer-Verlag, Berlin, Heidelberg, New York, Tokyo, 1984).

3) H. Sakaguchi, Prog. Theor. Phys. **80** (1988) 743.

4) C. Grebogi, E. Ott, and J. A. Yorke, Phys. Rev. Lett. **48** (1982) 1507.

5) C. Grebogi, E. Ott, and J. A. Yorke, Physica D **7** (1983) 181.

6) C. Grebogi, E. Ott, and J. A. Yorke, Phys. Rev. Lett. **57** (1986) 1284.

7) L. P. Shil'nikov, Dokl. Akad. Nauk SSSR **160** (1965) 558.

8) F. Argoul, A. Arneodo, and P. Richetti, Phys. Lett. A **120** (1987) 269.

9) F. T. Arecchi, W. Gadomski, A. Lapucci, H. Mancini, R. Meucci, and J. A. Roversi, J. Opt. Soc. Am. B **5** (1988) 1153.

REFERENCES

10) J. E. Ripper and T. L. Paoli, Appl. Phys. Lett. **17** (1970) 371.
11) K. Otsuka, IEEE J. Quantum Electron. **QE-13** (1977) 895.
12) K. Otsuka, Electron. Lett. **19** (1983) 723.
13) For a recent review, see D. Botez and D. E. Ackley, *IEEE Circuits Devices Mag.* **2** (1986) 8.
14) S. S. Wang and H. G. Winful, Appl. Phys. Lett. **52** (1988) 1774; H. G. Winful and S. S. Wang, Appl. Phys. Lett. **53** (1988) 1894.
15) Y. Aizawa, Prog. Theor. Phys. **81** (1989) 249.
16) H. Haken and M. Pauthier, IEEE J. Quantum Electron. **QE-4** (1968) 454.
17) S. J. Bhatt and C. S. Hsu, Trans. ASME J. Appl. Mech. **33** (1966) 113.
18) D. Pieroux, T. Erneux and K. Otsuka, Phys. Rev. A **50** (1994) 1822.
19) M. Toda, J. Phys. Soc. Jpn. **22** (1967) 431.
20) W. Kilische, H. R. Telle, and C. O. Weiss, Opt. Lett. **9** (1984) 561.
21) J.-L. Chern, J. K. McIver, and J.-T. Shy, Opt. Commun. **76** (1990) 63.

Chapter 6

COOPERATIVE DYNAMICS AND FUNCTIONS IN COLLECTIVE NONLINEAR OPTICAL ELEMENT SYSTEMS

Collective behavior, which arises through the interaction of elemental components, is currently of interest in nonlinear dynamics. Whether these components are neurons, amino acids, an so on, the collective behavior of the whole is qualitatively different from the sum of the individual parts. This concept is referred to as a *Gestalt* from the German philosophical concept that *the whole is more than the sum of its part*. Such collective systems are expected to result in automatic parallel processing and intelligent behavior. For example, neural networks and their optical analogue, that is, holographic systems, can function as associative memory [1–2]. The study of collective behaviors in coupled lasers and laser arrays are described in the preceding chapters. In this chapter, cooperative dynamics and functions of optical systems consisting of distributed nonlinear passive optical elements are described as a prototypical example of emergence of collective functions.

So far, we investigated the *temporally* irregular phenomena consiting of nothing more than the intrinsic temporal chaos inherent in the rules of their evolutionary process. In other terms, the discussions have mainly been restricted to *temporal* behaviors. A state of the nonlinear optical system is specified by the spatial configuration of the electric field which passes through the system as well. With the introduction of proper nonlinear evolutionary rule, it is not unusual that a spatially varying electric field can possess *spatially* unstable patterns (chaos) as stationary solutions. Especially, in the spatially coupled nonlinear element systems, spatial arrangements of stationary solutions and their dynamics become essentially important.

The fundamental question then arises as to whether it is possible to understand spatially irregular structures in terms of the intrinsic chaos inherent in the rule which determines their spatial arrangements. Underlying the question is the general problem of spatial disorder (chaos), an important subjects in the field of nolinear dynamics in nonequilibrium systems as well as in classical condensed matter.

Before duscussing spatiotemporal dynamics in spatially coupled nonlinear optical element systems in section **6.2**, let me describe spatial chaos in spatially extended *continuum* systems at first in section **6.1** as prototypes of spatial chaos in optical systems [A]. In section **6.2**, spatiotemporal dynamics in coupled nonlinear optical element systems consisting of multiple bistable elements are described [B]. In section **6.3**, applicability of cooperative dynamics in coupled nonlinear optical element systems is discussed, with the special attention being paid to the collective functions, which can hardly be expected in individual elements [B], [C].

6.1 Nonlinear Polarization Dynamics and Spatial Chaos

6.1.1 Polarization dynamics in crystal optics

An intense electromagnetic wave propagating in a nonlinear $\chi^{(3)}$ medium exhibits various self-action including self-focusing, self phase modulation etc. The self-induced birefringence effect is one of the interesting self-actions. In isotropic media including gas, liquid and glass, this effect has long been known as "ellipse rotation" [3]. This phenomenon is easily understood as follows: Let us consider the steady-state for brevity. In isotropic media, nonlinear polarization $\mathbf{P^{NL}}$ which is created by an electric field $\mathbf{E}$ is expressed by [3]

$$\mathbf{P^{NL}} = A\mathbf{E}(\mathbf{E}^* \cdot \mathbf{E}) + (1/2)B\mathbf{E}^*(\mathbf{E} \cdot \mathbf{E}), \tag{6.1}$$

where A, B are constant and depend on the physical origin of $\chi^{(3)}$. If (6.1) is rewritten in terms of circular polarizations, the following equation is obtained.

$$P_\pm^{NL} = [AE_\pm E_\pm^* + (A+B)E_\mp E_\mp^*]E_\pm, \tag{6.2}$$

where $E_\pm$ are clock-(anticlock-)wise rotating circularly-polarized field amplitudes. In general, A is smaller than $A+B$. As a result, if the elliptically-polarized electric field is passed through isotropic media, two circularly-polarized field components, which compose the elliptically polarized wave, experiences different phase shift due to Maxwell equation, and self-induced rotation of polarization ellipse takes place.

This phenomenon was discovered by Maker, Turhune and Savage and gave the first evidence of light intensity dependent refractive index $n^{(2)}$ in materials [3]. From the viewpoint of nonlinear dynamics, this phenomenon is not so attractive since the system is described by a spatial harmonic

6.1. NONLINEAR POLARIZATION DYNAMICS AND SPATIAL CHAOS

oscillator and the elliptically polarized light only rotates without changing its shape. If the proper nonlinearity is introduced into the nonlinear media, however, the situation drastically changes. In fact, tensor property inherent in $\chi^{(3)}$ nonlinearity in crystals results in attractive nonlinear polarization dynamics.

A. Spatial evolution of polarization state of light

We consider the geometry shown in Fig. 6.1. The total complex field in the medium can be decomposed into its x and y Cartesian components as

$$\mathbf{E} = [\tilde{E}_x(z,t)\hat{\mathbf{x}} + \tilde{E}_y\hat{\mathbf{y}}] = \hat{\mathbf{x}} E_x \exp(jk_x z - i\omega t) + \hat{\mathbf{y}} E_y \exp(jk_y z - j\omega t), \quad (6.3)$$

where k_x, k_y are wavenumbers for x- and y-polarized light and expresses the linear birefringence of the material.

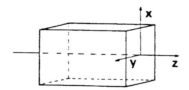

Figure 6.1: Anisotropic crystal referred to its principal direction.

The complex nonlinear polarization can similarly be represented as $\mathbf{P}^{NL} = P_x\hat{\mathbf{x}} + P_y\hat{\mathbf{y}}$, where each component of the polarization amplitude can be expressed in terms of a field-dependent suscesptibility tensor $\chi^{(3)} = \chi^{(3)}_{ijkl}$ and $\mathbf{P}^{NL} = \chi^{(3)}\mathbf{EEE}$. For a lossless Kerr medium, complex susceptibility χ is assumed to satisfy the Debye relaxation equation

$$\gamma \frac{\partial \chi}{\partial t} + \chi = \chi^{(1)} + \chi^{(3)} \; \mathbf{EE}, \quad (6.4)$$

where $\chi^{(1)} = \chi^{(1)}_{ij}$ (which is Hermitian), describes the linear birefringence corresponding to k_x and k_y.

The electric field and the nonlinear polarization are substituted into Maxwell equation, and the slowing-envelope approximation is made to yield the following equation

$$-j(\frac{\partial}{\partial z} + \frac{1}{c}\frac{\partial}{\partial t})E_i = k_i E_i + (2\pi\omega^2/k_i c^2)P_i^{NL}, \quad i = 1, 2, \quad (6.5)$$

where transformations of $x \to 1$ and $y \to 2$ are made. Equations (6.4) and (6.5) are fundamental equations which describe the spatiotemporal dynamics in lossless Kerr-like crystal.

In general, the relaxation rate γ might be index $(ijkl)$-dependent and therefore the exact spatiotemporal evolution is expected to be very complicated. Consequently, we restrict our discussion to the situation where the time response of the susceptibility is so fast that the time-varying terms of Debye's equation can be neglected. We further choose uniaxial crystals so that the section with (x, y) plane of the index ellipsoid is circular. Therefore the crystal has no birefringence, but we allow for a possible anisotropy in the third-order susceptibility tensor. In addition, for brevity, we assume parity-invariant crystals belonging to C_n ($n = 3, 4, 6$) symmetries. If we assume such crystals, including $3m$, $-3m$, 32 (trigonal), 422, $4mm$, $4/mmm$, $-42m$ (tetragonal), 622, $6mm$, $6/mmm$, $-6m2$ (hexagonal), 432, $-43m$, $m3m$ (cubic) [4], Eq. (6.5) can be written as

$$-j\frac{dE_i}{dz} = kE_i + E_i(\chi_1|E_i|^2 + 2\chi_2|E_{(3-i)}|^2)$$
$$+\chi_3 E_i^* E_{(3-i)} E_{(3-i)}, \quad i = 1, 2, \quad (6.6)$$

where
$$\chi_1 = (2\pi k_0^2/k)\chi_{1111}^{(3)}(\omega;\omega,\omega,-\omega), \quad (6.6a)$$

$$\chi_2 = (\pi k_0^2/k)[\chi_{1122}^{(3)}(\omega;\omega,\omega,-\omega) + \chi_{1212}^{(3)}(\omega;\omega,\omega,-\omega)], \quad (6.6b)$$

$$\chi_3 = (2\pi k_0^2/k)\chi_{1221}^{(3)}(\omega;\omega,\omega,-\omega), \quad (6.6c)$$

and $\chi_{ijkl}^{(3)}(\omega;\omega,\omega,-\omega)$ is stricktly real. For isotropic media, $\chi_3 + 2\chi_2 - \chi_1 = 0$ ($A = \chi_1 + \chi_2$, and $B = 2\chi_3$ in Eqs. (6.1) and (6.2)). Equations (6.6, 6.6a, 6.6b, 6.6c) are fundamental equations for describing the spatial evolution of polarization state in the crystal.

B. Formation of Duffing's oscillatory patterns

For a coherent beam there are three independent parameters called Stokes parameters describing the polarization state of light, which are bilinear products of the vector components of the complex field amplitude. They are real and contain all the information contained in that amplitude except the absolute phase of the field, which is usually not of interest. Therefore, we employ Stokes vector representation of the dynamical equation (6.6) [5]. Stokes parameters are written as

$$S_0 = |E_1|^2 + |E_2|^2, \quad (6.7a)$$

$$S_1 = |E_1|^2 - |E_2|^2, \quad (6.7b)$$

$$S_2 = E_1 E_2^* + E_1^* E_2, \quad (6.7c)$$

$$S_3 = j(E_1^* E_2 - E_1 E_2^*), \quad (6.7d)$$

6.1. NONLINEAR POLARIZATION DYNAMICS AND SPATIAL CHAOS

where $S_0^2 = S_1^2 + S_2^2 + S_3^2$. If we introduce these parameters, Eq. (6.6) can be written in the compact form

$$\dot{S}_0 = 0, \tag{6.8a}$$

$$\dot{\mathbf{S}} = \mathbf{\Omega_{NL}(S)} \times \mathbf{S}, \tag{6.8b}$$

where a dot stands for d/dz. The dynamics of the second equation is equivalent to the dynamics of the total angular momentum of the torque-free motion of a rigid body. The motion of the state of polarization as represented by the *reduced* Stokes vector $\mathbf{S} \equiv (S_1, S_2, S_3)$ is a non-rigid rotation of the Poincaré sphere S_0^2 with angular velocity

$$\mathbf{\Omega_{NL}(S)} \equiv \chi(0, (1-\lambda)S_2, -\lambda S_3), \tag{6.9}$$

where $\chi \equiv 2\chi_3$ and $\lambda = (\chi_1 - 2\chi_2 + \chi_3)/\chi_3$.

Equation (6.8) is integrable as follows. The invariants of the system are

$$S_1^2 + S_2^2/\lambda = \Gamma, \tag{6.10}$$

$$S_1^2 + S_2^2 + S_3^2 = S_0^2 \equiv R^2. \tag{6.11}$$

The intersections of the above surfaces are the closed trajectories along which the vector $\mathbf{S}$ moves. The nonlinear equation (6.8) is separable; S_1 satisfies the following equation of motion

$$\ddot{S}_1 + cS_1 + aS_1^3 = 0, \tag{6.12}$$

where $c = \chi^2 \lambda [R^2 + (1-\lambda)\Gamma]$, $a = 2\chi^2 \lambda (\lambda - 1)$. Equation (6.12) is nothing more than an anharmonic undamped Duffing oscillator. Therefore, if such solutions are dynamically stable, Duffing's oscillatory patterns are formed in the crystal as spatial structures.

Equation (6.12) is solved in terms of Jacobian elliptic functions as

$$S_1(z) = S_{10} cn(z(c + aS_{10}^2)^{1/2}, m),$$
$$m^2 = aS_{10}^2/2(c + aS_{10}^2), \quad for \ H \geq 0, \tag{6.13a}$$

$$S_1(z) = S_{10} dn(z(a/2)^{1/2} S_{10}, n),$$
$$n^2 = 2(1 + c/aS_{10}^2), \quad for \ H < 0, \tag{6.13b}$$

where $H = (1/2)\chi^2 \lambda \Gamma (R^2 - \lambda \Gamma)$ is the *energy* of the oscillator and for simplicity we chose the origin of the z axis such that $S_{20} = 0$ results. From invariants (Eqs. (6.10), (6.11)) and Eq. (6.13), the resulting trajectories on the Poincaré sphere are given as Fig. 6.2(b). As shown in Eq. (6.12), S_1 moves in a quartic potential well

$$V(S_1) = (1/2)cS_1^2 + (1/4)aS_1^4. \qquad (6.14)$$

Each trajectory on the sphere fixes the energy and shape of the potential. It is found from Fig. 6.2(b), the system has four stable singular points (center) located at potential minima $(\pm R, 0, 0)$ and $(0, 0, \pm R)$ and two unstable singular points (saddle) at potential maxima $(0, \pm R, 0)$. These eigenpolarizations satisfy $\mathbf{\Omega_{NL}}(\hat{\mathbf{S}}) \times \hat{\mathbf{S}} = 0$, i.e., $\hat{S}_1 \hat{S}_2 = 0$, $\hat{S}_2 \hat{S}_3 = 0$ and $\hat{S}_3 \hat{S}_1 = 0$. The former singular points correspond to circular eigenpolarizations which do not change their states during the propagation. Such trajectories featuring a separatrix orbits play an essential role to create chaotic spatial pattern as will be discussed later.

For isotropic media, we have $\lambda = 1$ (i.e., $a = 0$) and $\mathbf{\Omega_{NL}}(\mathbf{S}) = \chi(0, 0, -S_3)$. As a result, saddles disappear and the system exhibits simple ellipse rotation as shown in Fig. 6.2(a).

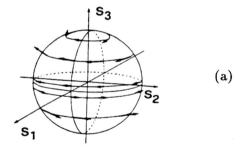

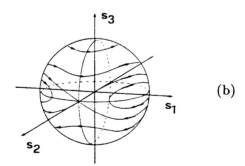

Figure 6.2: Trajectories of the polarization state on the Poincaré sphere. (a) isotropic media (b) anisotropic crystal.

6.1. NONLINEAR POLARIZATION DYNAMICS AND SPATIAL CHAOS

So far we have described the polarization dynamics in materials without a linear anisotropy. However, the introduction of linear anisotropy results in the spatial bifurcation phenomena of trajectories due to the interplay between linear birefringence and self-induced effect. Here, we discuss such effect briefly. If we consider isotropic media with a linear anisotropy such as birefringent optical fibers, the elementary resultant motion is the sum of the two separate phenomena. These are a linear birefringence with angular velocity $\mathbf{\Omega_L} = (\eta, 0, 0)$ and a self-induced ellipse rotation with angular velocity $\mathbf{\Omega_{NL}(S)} = \chi(0, 0, -S_3)$, where $\eta \equiv k_1 - k_2$, and Eq. (6.8b) is modified to

$$\dot{\mathbf{S}} = \mathbf{\Omega(S)} \times \mathbf{S} = [\mathbf{\Omega_L} + \mathbf{\Omega_{NL}(S)}] \times \mathbf{S}, \tag{6.15}$$

In this case S_1 satisfies the equation

$$\ddot{S}_1 + d + cS_1 + bS_1^2 = 0, \tag{6.16}$$

where $c = \chi^2(R^2 - \Gamma') + 4\eta^2$, $b = 3\eta\chi$, $d = \chi\eta(R^2 - 2\Gamma')$, and $\Gamma' = S_1^2 + S_2^2 + 2\eta S_1/\chi$ is an invariant. The solution of Eq. (6.16) can be expressed through the elliptic integral

$$z = \int_{S_{10}}^{S_1} Q(x)^{-1/2} dx \tag{6.17}$$

where $Q(x) = H' - dx - (1/2)cx^2 - (1/3)bx^3$ and $H' = (1/2)\chi^2\Gamma'(R^2 - \Gamma')$.

A stability analysis of the singular points of Eq. (6.15) reveals how the interaction between the linear birefringence and the ellipse rotation takes place. The eigenpolarizations $\hat{\mathbf{S}}$ satisfies $\mathbf{\Omega(\hat{S})} \times \hat{\mathbf{S}} = 0$, i.e.,

$$\hat{S}_1\hat{S}_3 + p\hat{S}_3 = 0, \quad \hat{S}_2 = 0, \tag{6.18}$$

where $p = \eta/\chi$ is a bifurcation parameter. From Eq. (6.17), the system admits the two stable singular points (circular eigenpolarizations) at $(\pm R, 0, 0)$ in the low field intensity regime $R < p$ (see Fig. 6.3(a)). In this regime, the linear birefringence dominates and the trajectory is topologically equivalent to that of a rigid rotation around S_1 and the dynamics reduces to a simple Bloch equation with a constant magnetic field. If the field intensity is increased such that $R > p$, the separatrix orbit appears similarly to the case of anisotropic crystals and the point $(-R, 0, 0)$ bifurcates into an unstable saddle point at $(-R, 0, 0)$ and two centers (circular eigenpolarizations) at $(-p, 0, \pm(R^2 - p^2)^{1/2})$ as shown in Fig. 6.3(b).

6.1.2 Spatial chaos in polarization

As discussed in Chapters **2, 3, 4**, complex behavior, e.g. chaos, is expected in general in the situation that nonlinear oscillators such as lasers

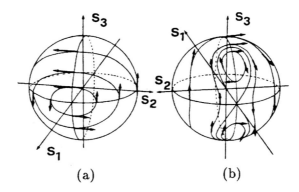

Figure 6.3: Representation of the field polarizations on the Poincaré sphere of a birefringent optical fiber. (a) $R < p$, (b) $R > p$.

are perturbed periodically by an external force at frequencies in the neighborhood of the fundamental oscillation or one of the harmonics or these oscillators are coupled to each other such as laser arrays. The simplest example is a pendlum driven by periodic force or pendlums coupled nonlinearly to each other.

In the previous subsection, we described spatial anharmonic oscillatory solutions expected in nonlinear polarization dynamics. These anharmonic undamped Duffing oscillators, featuring separatrices, are good candidates for realizing spatial chaos by adding other degrees of freedom such as a periodic perturbation or a nonlinear coupling.

A. *Unidirectional coupling - birefringent fiber with periodic coupling* [6-7]

As discussed in the previous subsection, the polarization trajectories of an unperturbed birefringent optical fiber above threshold, i.e. $R > p$, are renpresented by separatrix orbits as shown in Fig. 6.3(b). If such an undamped nonlinear oscillator is perturbed by external periodic force, the onset of stochasticity is expected. Therefore, let us introduce here a periodic mode coupling into a birefringent fiber and examine its dynamics numerically.

The periodic mode coupling is introduced by means of a periodic distribution of stress, Faraday rotations, periodic rocking of the preform while drawing and uniform twisting while being wound onto a drum. If such periodic coupling is introduced, the linear birefringence vector $\mathbf{\Omega_L}$ is modified to $(\eta, \eta_p \cos(\omega_p z + \phi_0), 0)$, where η_p is the perturbing birefringence for simplicity oriented at 45° to fiber axis and ω_p is the spatial modulation frequency. A nonlinear resonance occurs whenever the input beam is such that the spatial modulation frequency is matched with the fundamental os-

6.1. NONLINEAR POLARIZATION DYNAMICS AND SPATIAL CHAOS

cillation or one of the harmonics. Moreover, the accumulation of resonances in a thin layer near the separatrix destroys the regularity of the motion and spatial chaos develops. Figure 6.4 depicts the numerically obtained discrete Poincaré map of **S** on the southern hemisphere ($S_1 < 0$). (a) shows periodic or quasiperiodic evolution for different input polarizations. With the increase in wave intensity, i.e. $R > p$, stochastic motion takes place near separatrix of the unperturbed system as shown in (b). A abrupt transition from regular to irregular evolution occurs by the variation of input polarization across the Kolomogorov-Arnol'd-Moser (KAM) tori bounding the chaotic domain.

B. Bidirectional coupling - degenerate four-wave mixing [A], [5]

In the previous system, the interaction is unidirectional and there is no

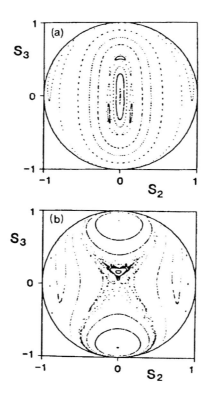

Figure 6.4: Poincaré map produced of S on the southern hemisphere of S^2. (a) $R/p = 1$ (b) $R/p = 2$ with $\eta_p/\eta = 0.01$. Three polarization islands correspond to fundamental resonance and the elliptic fixed points at the center of the island denote a stable period-3 solution of (a).

boundary condition for determining the polarization state of light. Therefore, the dynamics are relatively simple. Here, we discuss the bidirectional coupling scheme. When two counterpropagating light beams interact in nonlinear optical media (degenerate four-wave mixing scheme) as shown in Fig. 6.5, complex polarization dynamics is also expected due to the nonlinear coupling of two anharmonic Duffing oscillators possessing separatrices. Yumoto and Otsuka pointed out the possible existence of "spatial chaos" for the first time by introducing this scheme and initiated the research of nonlinear polarization dynamics [A]. In this scheme, Eq. (6.8) is modified into the following coupled equations.

$$\dot{S}_{0,F} = 0, \quad \dot{\mathbf{S}}_{\mathbf{F}} = [\mathbf{\Omega}_{\mathbf{NL}}(\mathbf{S}_{\mathbf{F}}) + \mathbf{\Omega}'_{\mathbf{NL}}(\mathbf{S}_{\mathbf{B}})] \times \mathbf{S}_{\mathbf{F}}, \qquad (6.19)$$

$$\dot{S}_{0,B} = 0, \quad -\dot{\mathbf{S}}_{\mathbf{B}} = [\mathbf{\Omega}_{\mathbf{NL}}(\mathbf{S}_{\mathbf{B}}) + \mathbf{\Omega}'_{\mathbf{NL}}(\mathbf{S}_{\mathbf{F}})] \times \mathbf{S}_{\mathbf{B}}, \qquad (6.20)$$

where subscripts F and B correspond to forward- and backward-traveling waves and $\mathbf{\Omega}'_{\mathbf{NL}}(\mathbf{S}) \equiv ([\chi' + \chi(2\lambda - 1)]S_1, (\chi' + \chi)S_2, (\chi' - \chi)S_3)$ and $\chi' \equiv 2\chi_2$. $\mathbf{\Omega}_{\mathbf{NL}}$ represents the self-induced birefringence, whereas $\mathbf{\Omega}'_{\mathbf{NL}}$ is the mutually induced birefringence between the two counterpropagating waves and includes the refractive index grating effect which results in mutual scattering of two waves (local feedback effect).

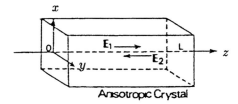

Figure 6.5: Configuration of degenerate four-wave-mixing scheme.

Tratnik and Sipe carried out an intensive analysis for general symmetry classes. They proved that for propagation along an axis with C_n ($n = 1, 2, 4$) rotational symmetry, there are only three invariants, including two intensities $S_{0,F}^2$, $S_{0,B}^2$ and the free-energy, where the system has six degrees of freedom [8]. This means that chaotic behavior is possible according to a theorem from analytical dynamics [9], which concerns n degrees of freedom systems with $n - 2$ invariants. They also showed that as for C_3, C_6 or C_∞ (*isotropic*) symmetry, there are two additional invariants associated with effective rotational invariance about the propagation axis at the level of a $\chi^{(3)}$ nonlinearity. Consequently, these systems cannot exhibit chaos.

Therefore, we restrict the following discussions to C_4 symmetry classes. We employ here a parity-invariant crystal KTa$_{0.65}$Nb$_{0.35}$O$_3$ (KTN) and

6.1. NONLINEAR POLARIZATION DYNAMICS AND SPATIAL CHAOS

numerical simulations of Eqs. (6.19) and (6.20) are carried out for different initial polarization configurations at $z = 0$. Solutions can be categorized as follows:

(1) Nonlinear eigenpolarizations: The beams pass through the crystal without changing their polarization state. They are obtained from $\mathbf{S_F} = \mathbf{S_B} = 0$ and correspond to fixed points of the system. The corotating and counter-rotating circular polarizations are examples of nonlinear eigenpolarizations.

(2) Resonance state: The polarizations states of two beams are synchronized and change periodically in space. They are obtained by setting $\mathbf{S_F} = \mathbf{S_B}$ in Eqs. (6.19) and (6.20) and can be expressed analytically in terms of elliptic functions similarly to the unidirectional case.

(3) Multiperiodic solutions.

(4) Quasiperiodic solutions.

(5) Chaotic solutions.

Cases (2)–(5) are depicted in Fig. 6.6. In these illustrations, the polarization direction, i.e. $\theta_p = E_{px}/E_{py}$ ($p = 1, 2$), is shown as a function of z for visualizing the spatial evolution of polarization. Here, $p = 1(2)$ denotes the wave propagating toward $+z(-z)$ direction. The bifurcation

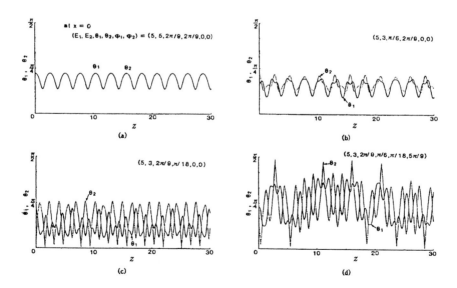

Figure 6.6: Steady-state spatial evolutions of polarization state of light for different initial conditions in degenerate four-wave mixing scheme. Initial conditions are shown in each figure, where $\phi_p = \phi_{py} - \phi_{pz}$ ($p = 1, 2$) are nonlinear phase terms. (a) period-1, (b) period-6, (c) quasiperiodic, (d) chaotic. In the figure, polarization state is exppressed by $\theta_p = E_{px}/E_{py}$, where E_{px} and E_{py} ($p = 1, 2$) are field amplitudes along crystal axes in Fig. 6.5.

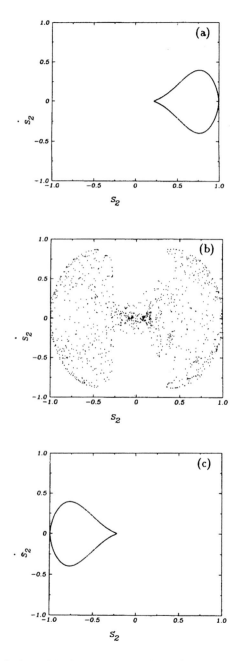

Figure 6.7: Poincaré plots of Stokes parameters for a C_4 parity-invariant crystal, e.g., KTN in degenerate four-wave mixing scheme.

scenario of the present system has yet to be known, however, the quasiperiodic route to chaos has been shown for linear polarization configurations by Tratnik and Sipe [10]. Poincaré plots for different values of ϕ are shown in Fig. 6.7, where the polarization direction of $\mathbf{S_F}$ (linear polarization) rotated with respect to $\mathbf{S_B}$ (linear polarization) by an angle of $\phi/2$. A number of such calculations indicate quasiperiodic behavior for ϕ less than about 141.0°. Figure 6.7(a) indicates two-frequency motion. When ϕ reaches about 141.35°, an abrupt transition to chaos occurs as shown in Fig. 6.7(b) and this chaotic regime then seems to persist to about 218.6°. Above this, an abrupt transition back to regular behavior as shown in Fig. 6.7(c) takes place.

On the contrary to the unidirectional coupling scheme in birefringent optical fibers, the local feedback mechanism stemming from refractive grating is essential for degenerate four-wave mixing scheme. This results in multiple solutions if appropriate boundary conditions are imposed similary to the usual multistable devices with feedback. Indeed, polarization bi-(multi-)stability has been predicted [11] and been demonstrated [12] for degenerate four-wave mixing schemes.

6.2 Spatiotemporal Dynamics of Coupled Nonlinear Optical Element Systems

If chaos can be generated as spatial patterns instead of temporal patterns as described in the previous section, *spatial chaos*, or more strictly speaking a spatially unstable periodic pattern embedded in the chaotic patterns, seems likely to be temporally stable. Such a possibility has been pointed out by Aubry in a completely different physical context [13]. Aubry discussed the Frenkel-Kontrova system, which is a *microscopic equilibrium system*, and showed that a large variety of spatially unstable periodic solutions embedded in the chaotic sea are dynamically stable at zero temperature. At a finite temperature, however, these solutions become metastable since the system can stepover the barrier of local minima thermally.

The nonlinear polarization dynamics discussed in the previous section suggests the possibility of the existence of spatial chaos in *macroscopic nonequilibrium systems*, however, the dynamic analysis of these systems is very difficult and no information has been obtained regarding its dynamical stability.

Otsuka and Ikeda predicted the existence of dynamically stable spatial chaos in the proposed coupled nonlinear optical element system [B]. Firth also predicted spatial chaos in coupled bistable pixels [14]. In this section, spatial chaos and spatiotemporal dynamics in these proposed macroscopic nonequilbrium systems are presented.

6.2.1 Otsuka-Ikeda model

In the conceptual model shown in Fig. 6.8, nonlinear elements possessing third-order susceptibility are arranged in an optical ring cavity. These elements interact via counterpropagating light beams $(A_F^{(k)}, A_B^{(k)})$ that are introduced through the mirrors which separate the elements. The forward and backward propagating waves couple through the nonlinear refractive index grating formed in the medium. This effect makes the analysis quite difficult. Therefore, we assume that the refractive index grating diffuses very fast and we neglect the mutual coupling through the refractive index grating. Then the Maxwell-Debye (or Bloch) equation which describes the motion of electric fields $E_F^{(k)}$, $E_B^{(k)}$ and the nonlinear phase shift introduced into the electric field at each cell $\phi_F^{(k)}$, $\phi_B^{(k)}$ can be reduced to the following delay differential equations in the *dispersive limit* by adding the term coming from $E_B^{(k)}$ to Eqs. (1.49)–(1.50) in Chapter **1**:

$$E_F^{(k)}(t+t_R) = A_F^{(k)} + B\exp(i(\phi_F^{(k-1)}(t) + \phi_0^{(k-1)}))E_F^{(k-1)}(t), \qquad (6.21)$$

$$E_B^{(k)}(t+t_R) = A_B^{(k)} + B\exp(i(\phi_B^{(k+1)}(t) + \phi_0^{(k+1)}))E_B^{(k+1)}(t), \qquad (6.22)$$

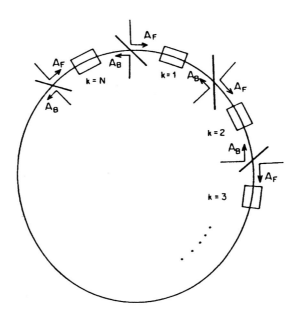

Figure 6.8: Conceptual model of a nonlinear optical system with distributed nonlinear elements.

6.2. SPATIOTEMPORAL DYNAMICS

$$\tau(\partial \phi_F^{(k)}(t)/\partial t) = -\phi_F^{(k)}(t) + q[(1-e^{-\alpha\ell})/\alpha]|E_F^{(k)}(t)|^2$$
$$+ q\int_0^\ell ds\, e^{-\alpha s}|E_B^{(k)}(t+\ell/c - 2s/c)|^2, \quad (6.23)$$

$$\tau(\partial \phi_B^{(k)}(t)/\partial t) = -\phi_B^{(k)}(t) + q[(1-e^{-\alpha\ell})/\alpha]|E_B^{(k)}(t)|^2$$
$$+ q\int_0^\ell ds\, e^{-\alpha s}|E_F^{(k)}(t+\ell/c - 2s/c)|^2. \quad (6.24)$$

Here, $\tau = 1/\gamma$, c is the velocity of light, $B = \sqrt{R}e^{-\alpha\ell/2}$ is the coupling coefficent between the adjacent cells (R: mirror reflectivity, α: absorption coefficient, ℓ: cell length), $\phi_0^{(k)}$ is the linear phase shift across the k-th cell and $q \equiv \text{sign}(n_2)[\alpha/(1-e^{-\alpha\ell})]$ (n_2: quadratic coefficient of the nonlinear refractive index).

To extract the essence of a coupled element system, let us simplify these equations further. First, we assume that the variation of the electric field intensity during the transit time through the cell is negligible, i.e., $t_R = \ell/c \to 0$ and that absorption by the medium is negligible, i.e., $\alpha\ell \to 0$.

Next, we consider the limit of large dissipation, i.e., $B = \sqrt{R} \ll 1$ with $A_{F,B}^{(k)2}B \sim O(1)$. Then, Eqs. (6.21) and (6.24) reduce to:

$$\tau\dot\phi_k = -\phi_k + f_F(\phi_{k-1}) + f_B(\phi_{k+1}), \quad (6.25)$$

with

$$f_{F,B}(\phi_k) \equiv A_{F,B}^{(k)2}[1 + 2B\cos(\phi_k + \phi_0)] = |E_{F,B}^{(k)}|^2, \quad (6.26)$$

Here, the input fields as well as the linear phase shift do not depend on the cell, i.e., $A_F^{(k)} = A_F$, $A_B^{(k)} = A_B$, $\phi_0^{(k)} = \phi_0$. In such an ideal case, the state of each cell is represented by the nonlinear phase shift and the spatial pattern of ϕ_k becomes important.

A. Unidirectional coupling

First let us consider the case of unidirectional coupling, i.e., $A_F^{(k)} = A$ and $A_B^{(k)} = 0$ and examine the dynamic stability of stationary spatial structures. In this case, the stationary solution, $\bar\phi_k$ of Eq. (6.25), is determined by the mapping rule

$$\bar\phi_{k+1} = f_F(\bar\phi_k). \quad (6.27)$$

The properties of the solutions of mapping rule Eq. (6.27) is shown in **1.5.2**. In particular, the solution is shown to exhibit complicated bifurcation phenomena when the nonlinear parameter, that is A^2 in our case, is changed. The spatial structures which can exist in our case are those whose spatial period is N or its divisor that satisfies the following boundary condition

$$\bar{\phi}_{N+1} = \bar{\phi}_1. \tag{6.28}$$

However, let us forget this restriction for a moment and look at the bifurcation phenomena based on Eq. (6.27). This corresponds to the case of an open ring without feedback and the situation may be very much similar to the case of periodically perturbed birefringent fibers discussed in the previous section.

The solutions which are obtained by iterating the map Eq. (6.27) are shown in Fig. 6.9 assuming $B = 0.1$ and $\phi_0 = 0$. From this figure, the global behavior can be understood. In short, the stable period-one cycle solution exhibits successive *spatial period-doubling bifurcations* leading to *spatial chaos*. Within the chaotic regime, a lot of stable periodic solutions whose period p are embedded as windows.

What happens when the ring is closed and the boundary condition Eq. (6.28) is introduced? Among these solutions, spatial solutions which satisfy the boundary condition of Eq. (6.28) and are dynamically stable can exist as stable structures. The dynamical stability of these solutions can be examined by the linear stability analysis of Eqs. (6.25) and (6.26). The spectrum of the small deviation around the stationary solutions $\delta\phi_k = \delta\tilde{\phi}_k \exp(\lambda t)$ is governed by the following characteristic equation:

$$(\lambda + 1)^N = \exp(N\alpha) \operatorname{sgn} \sigma, \quad \sigma = \prod_{k=1}^{N} f'_F(\bar{\phi}_k), \tag{6.29}$$

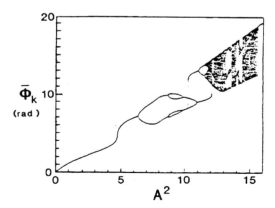

Figure 6.9: Bifurcation diagram for $\phi_0 = 0$ and $B = 0.1$.

where

$$\alpha = (1/N) \sum_{k=1}^{N} \ln |f'_F(\bar{\phi}_k)| \qquad (6.30)$$

is the "spatial Lyapunov exponent" of the periodic solutions of $\bar{\phi}_1, \bar{\phi}_2, \ldots, \bar{\phi}_N$. Obviously, the structures whose $\alpha < 0$ correspond to the "stable" periodic solutions of Eq. (6.27) and the structures whose $\alpha > 0$ correspond to "unstable" periodic solutions of Eq. (6.27). The structures of $\alpha < 0 (> 0)$ are referred to as *spatially* stable (unstable) structures. Equation (6.29) indicates that the *dynamic* stability of the structures, which is evaluated by the sign of the real component of $N\lambda's$, are closely related to the *spatial* stability through the spatial Lyapunov exponent. From the above analysis, it is easy to show the following result:

The spatially stable structures are dynamically stable, while the spatially unstable structures, e.g., spatial chaos, are dynamically unstable

In short, in unidirectional coupling, the realizable solutions are almost spatially stable, thus they are not spatial chaos. The spatial chaos is dynamically unstable and is converted into spatiotemporal chaos in the unidirectional coupling scheme. In order to confirm this result, the ϕ_k are plotted, except for transients, as $P = A_F^2 + A_B^2$ is increased (or decreased) very slowly in time (see Fig. 6.10(a)) assuming $N = 23$. In this case, since N is a prime number the spatial period-doubling is inhibited and the spatiotemporal chaos (STC) develops almost everywhere excepting a trivial period-one solution.

A similar relationship between the dynamical stability and the spatial stability might be expected in the case that the periodically perturbed birefringent fiber is looped and the boundary condition is introduced.

B. Bidirectional coupling

Next, let us consider the case of bidirectional coupling where $A_F^{(k)} = A_B^{(k)} = A$. In this case, the dynamics are dramatically changed. Although the interactions between the nearest neighbor cells are symmetric, Eq. (6.25) lacks the potential condition of $\partial \dot{\phi}_k / \partial \phi_{k-1} = \partial \dot{\phi}_{k-1} / \partial \phi_k$, and thus there is no Lyapunov functional which corresponds to a Hamiltonian (or free energy) in equilibrium systems. Therefore, one cannot ensure an approach to static configurations. Indeed, STC takes place in a regime of quite high intensity (see Fig. 6.10(b)). In contrast to the unidirectional case, however, self-induced spatial disorder is found to be stabilized over wide regions as shown in Fig. 6.10(b). We see that ϕ_k varies stepwise with P, being accompanied by hysteresis, although the STC appears on the higher intensity side ($P \geq 20$). On the low intensity side of each step, ϕ_k is multifurcated into N different static values. This multifurcation implies

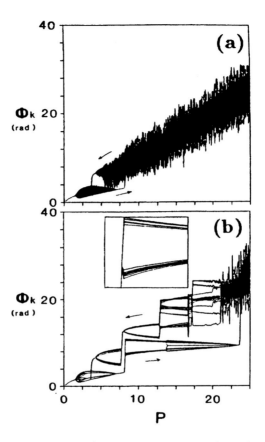

Figure 6.10: $\phi_k(t)(k = 1, 2, \ldots, N)$ versus P. $B = 0.3$, $\phi_0 = 0$ and $N = 23$. (a) Unidirectional, (b) bidirectional. Inset: Enlargement around $P = 10$. In this case, asymmetric period-N solutions are realized and "multifurcation" into 12 different states is seen.

that the spatial distribution of ϕ_k is frozen into a state with different values at different sites. Global structure of hysteresis as well as of bifurcation does not depend on the system sizes. The inset is an enlargement around the input intensity of $P = 10$. Two typical patterns of spatially multifurcated structure are depicted in Fig. 6.11. These are obtained for low P ($P = 3.2$) and high P ($P = 11.18$) regimes. In the low P side, the spatial structure looks almost like a period-two cycle pattern into which "defect-like" structures are inserted. In the high P side, the spatial structure is quite irregular, but it seems to wander between different period-two cycle structures. Therefore, let us then look at the stable domain of period-two

6.2. SPATIOTEMPORAL DYNAMICS

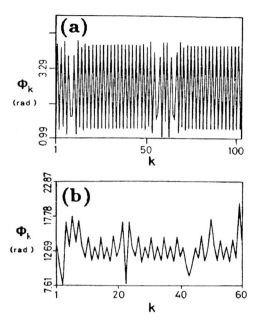

Figure 6.11: Typical patterns of spatial multifurcated structure in bidirectional coupling. (a) For the low-P regime: $P = 3.2$, $N = 103$, $B = 0.5$, $\phi_0 = 0$. (b) For the high-P regime: $P = 11.18$, $N = 60$, $B_0 = 0.5$, $\phi_0 = 0$.

cycle solutions which forms a basis for the "multifurcated" structure.

Stationary solutions for the bidirectional case are determined by the following rule, which corresponds to Eqs. (6.27) and (6.28) for the unidirectional case:

$$\bar{\phi}_k = f_F(\bar{\phi}_{k-1}) + f_B(\bar{\phi}_{k+1}), \quad \bar{\phi}_1 = \bar{\phi}_{N+1}. \tag{6.31}$$

This rule defines the following two-dimensional mapping rule, which determines $\bar{\Phi}_{k+1} \equiv (\bar{\phi}_{k+1}, \bar{\phi}_k)$ from $\bar{\Phi}_k \equiv (\bar{\phi}_k, \bar{\phi}_{k-1})$:

$$\mathbf{F} : (\bar{\phi}_k, \bar{\phi}_{k-1}) \longmapsto (\bar{\phi}_{k+1}, \bar{\phi}_k) = (f_B^{-1}(\bar{\phi}_k - f_F(\bar{\phi}_{k-1})), \bar{\phi}_k). \tag{6.32}$$

Here, "spatial instability" is used to imply that a positive Lyapunov exponent exists in the dynamic system along the spatial structure $(\bar{\phi}_1, \bar{\phi}_2, \ldots)$. On the other hand, the "dynamic stability" of the spatial structure $(\bar{\phi}_1, \bar{\phi}_2, \ldots)$ is determined by the sign of roots (λ) of the following linerazlied characteristic equation for the small deviation around the stationary solutions $\delta \phi_k = \delta \tilde{\phi}_k \exp(\lambda t)$:

$$\tau \lambda \delta \tilde{\phi}_k = -\delta \tilde{\phi}_k + f'_F(\bar{\phi}_{k-1}) \delta \tilde{\phi}_{k-1} + f'_B(\bar{\phi}_{k+1}) \delta \tilde{\phi}_{k+1}. \tag{6.33}$$

Here, we introduce the function

$$f_F(\phi) = f_B(\phi) \equiv (1/2)f_P(\phi). \tag{6.34}$$

When $P \equiv A_F^2 + A_B^2$ is small, the period-one cycle solution $\bar{\phi}_k = x_1$ satisfying $x_1 = f_P(x_1)$ is a stable structure. However, the period-one cycle structure becomes dynamically unstable and bifurcation takes place as P exceeds the critical value P_c. In this regime, the period-two cycle structure $\bar{\phi}_{2k-1}(\text{or } \bar{\phi}_{2k}) = x_1$, $\bar{\phi}_{2k}(\text{or } \bar{\phi}_{2k-1}) = x_2$, which is determined by $x_1 = f_P(x_2)$ and $x_2 = f_P(x_1)$, is realized similarly to the unidirectional case. The dynamic stability of these period-two cycle solutions is found to be greatly different from that for the unidirectional case. Let us examine the dynamic stability index λ for two coupling schemes, assuming N(cell number) is even. From a simple analysis of Eq. (6.33), the following equations are given:
(1) $S \equiv f'(x_1)f'(x_2) > 0$

$$\lambda = -1 + (1/2)|S|^{1/2}\cos(\theta/2) \quad (bidirectional), \tag{6.35}$$

$$\lambda = -1 + |S|^{1/2}\exp(i\theta/2) \quad (unidirectional). \tag{6.36}$$

(2) $S < 0$

$$\lambda = -1 + (1/2)|S|^{1/2}i\cos(\theta/2) \quad (bidirectional), \tag{6.37}$$

$$\lambda = -1 + |S|^{1/2}i\exp(i\theta/2) \quad (unidirectional), \tag{6.38}$$

where $\theta = 2k\pi/N (k = 0, 1, 2, \ldots, N)$.

As for the unidirectional case, the dynamically stable region ($\lambda < 0$) almost coincides with the region of $|S| < 1$ where the period-two cycle solutions are spatially stable. As a result, the region where the period-two cycle solutions are realized is extremely narrow as was described in the previous subsection A. In the bidirectional scheme, the situation drastically changes. In short, the period-two cycle structures become dynamically stable ($\lambda < 0$) even in the spatially unstable region; When P is increased, S first reaches 1 and the period-two cycle structure appears at this point. This structure is stable following Eq. (6.35). When P is increased further, S increases in a negative direction. The period-two cycle structure in the unidirectional case becomes unstable when S becomes smaller than -1 due to Eq. (6.38). In the bidirectional case, on the contrary, the period-two cycle structure is dynamically stable even when it is spatially destabilized, i.e., $S < -1$. (See Eq. (6.37).) This fact implies that the local feedback due to the bidirectional coupling much stabilizes the period-two cycle structure. Fig. 6.12 shows the stable domain of period-2 cycle solutions for bidirectional interaction. The period-2 structures are seen to be stabilized eveywhere excepting negative-slope branches.

6.2. SPATIOTEMPORAL DYNAMICS

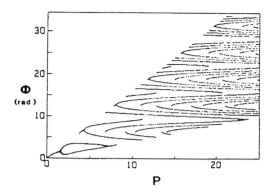

Figure 6.12: Multistable structure of period-2 cycle solutions as a function of P. $B = 0.3$ and $\phi_0 = 0$. Period-1 cycle solutions are also shown.

1. Frozen spatial heterostructure

The fact that the spatially unstable period-two cycle structures can exist stably in time is the key to the existence of spatial chaos. The period-two cycle structure is represented by two-dimensional vectors $\mathbf{x_2} = (x_1, x_2)$ and $\mathbf{x_2'} = (x_2, x_1)$, which are both the fixed points of the second iteration of the spatial evolution rule $\mathbf{\Phi}'' = \mathbf{F}(\mathbf{F}(\mathbf{\Phi}))$. The spatial instability means that the spatial evolution of a small deviation $\delta\mathbf{\Phi}(= \mathbf{\Phi} - \mathbf{x_2}$ or $\mathbf{\Phi} - \mathbf{x_2'})$ from the period-two fixed point (i.e., $\mathbf{\Phi} = \mathbf{x_2}$ or $\mathbf{x_2'}$) of the second iteration $\mathbf{\Phi}'' = \mathbf{F}(\mathbf{F}(\mathbf{\Phi}))$ separates from $\mathbf{x_2}$ or $\mathbf{x_2'}$ such that $\delta\phi_k \propto \exp(\nu^+ k)$. Here, the second iteration is determined from Eq. (6.32) and $\nu^+ (|\nu^+| > 1)$ is one of the two characteristic roots of the following variational equation:

$$\delta\mathbf{\Phi}'' = \left.\frac{\partial \mathbf{F}(\mathbf{F}(\mathbf{\Phi}))}{\partial \mathbf{\Phi}}\right|_{\mathbf{\Phi}=\mathbf{x_2}} \delta\mathbf{\Phi} \equiv \mathbf{G}\delta\mathbf{\Phi}. \quad (6.39)$$

The spatial instability (i.e., $|\nu^+| > 1$) takes place only for a real root. Since $\det \mathbf{G} = 1$, there is another root $\nu^- (|\nu^-| < 1)$. Therefore, $\mathbf{x_2}$ ($\mathbf{x_2'}$) is a hyperbolic fixed point accompanied by manifolds W_u and W_s passing through it (Fig. 6.13). Consider the motion of the vector $\mathbf{\Phi}$ governed by the second iteration $\mathbf{\Phi}'' = \mathbf{F}(\mathbf{F}(\mathbf{\Phi}))$. The vector $\mathbf{\Phi}$ approaches $\mathbf{x_2}$ ($\mathbf{x_2'}$) along W_s, whereas it separates from $\mathbf{x_2}$ ($\mathbf{x_2'}$) along W_u. In the vicinity of $\mathbf{x_2}$ ($\mathbf{x_2'}$), W_u and W_s are in the directions of eigenvectors $\mathbf{q}^\pm$ corresponding to the two roots $\mathbf{G}\mathbf{q}^\pm = \nu^\pm \mathbf{q}^\pm$. Manifolds W_u and W_s are called the unstable and stable manifolds. If W_u of $\mathbf{x_2}$ and W_s of $\mathbf{x_2'}$ cross transversally, a so-called heteroclinic orbit is produced and the orbit of the second iteration $\mathbf{\Phi}'' = \mathbf{F}(\mathbf{F}(\mathbf{\Phi}))$ near $\mathbf{x_2}$ and $\mathbf{x_2'}$ is expected to exhibit extremely complex (i.e., chaotic) behavior [15]. Moreover, since $\mathbf{x_2}$ and $\mathbf{x_2'}$ are dynamically stable, the spatial structure which corresponds to heteroclinic chaos is expected to

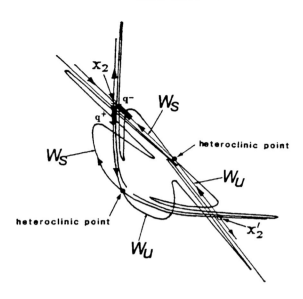

Figure 6.13: Schematic illustration of a heteroclinic crossing.

be stable in time.

If a heteroclinic crossing takes place in this system, what kind of spatial pattern does this heteroclinic orbit produce? A state vector Φ starts from the neighborhood of x_2, and separates from x_2 along W_u and switches to W_s, finally approaching x_2'. Since x_2' is a hyperbolic fixed point, Φ is repelled by x_2' and separates from x_2' along W_u. Thus the heteroclinic orbit indicates the existence of a spatial structure which switches between x_2 and x_2' chaotically in the manner $x_2 \to x_2' \to x_2 \cdots$. The switch from x_2 to x_2' corresponds to the "defect" shown in Fig. 6.11(a). The defect is considered to be the interface, i.e., "kink" which connects two types of period-two cycle structures. Figure 6.14 is an example of such a kink structure (a) and a spatial return map (b) obtained in the first multifurcated branch in Fig. 6.10(b). The heteroclinic closing is apparent from (b). From the numerical analysis, the temporal stability exponent λ is shown to be negative and coincide with the spatial Lyapunov exponent, yielding that these heteroclinic spatial chaos structures are dynamically stable.

Then, how high can we increase the density of kinks in the present system? From an extensive numerical simulation, the minimum distance between the kinks is found to be five lattice size, and the following rule is obtained.

Consider the α-structure $(+ - - + -)$ which contains a kink and is

6.2. SPATIOTEMPORAL DYNAMICS

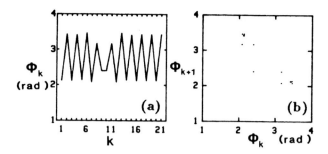

Figure 6.14: An example of "kink" structures for $N = 21$, $P = 5$, $B = 0.3$ and $\phi_0 = 0$. (a) spatial arrangement, (b) spatial return map.

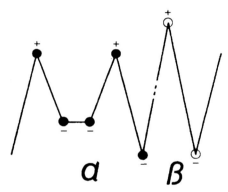

Figure 6.15: Fundamental structures which construct dynamically stable heteroclinic structures. $+--+-$: α structure, $+-$: β structure.

formed with 5 lattices sites and the β-structure $(+-)$ which is a fundamental unit of period-2 cycle structure as shown in Fig. 6.15. Structures which are obtained by arbitrary combination of α and β are dynamically stable

The kink structure α', which has the complementary binary sequence $(-++-+)$ is another stationary solution of the present system. However, as will be discussed in **6.3**, this type of kink forms a dynamically unstable fixed point (saddle) with a positive temporal stability exponent and cannot be realized.

2. Route to spatiotemporal chaos

The heteroclinic structures discussed so far are "interface" solutions switching between period-2 cycle solutions. However, spatial chaos structures are not restricted to this simple interface structures, In large P regions, chaotic patterns, which make rounds between an extremely large variety of different fundamental periodic structures, have been shown to coexist stably in time at a fixed P [B-2]. In a large-N system, an extremely large number of spatial chaos patterns like Fig. 6.11(b) appear. Therefore, we assume a small-N system ($N = 8$). Note that the fundamental structures including α and β have no meaning at all for such a small cell number. Figure 6.16 shows stable structures for $P = 15$, indicating five coexisting patterns. As P increases, the number of coexisting structures increases, and finally these solutions become unstable and are replaced by STC.

The question then arises:

What kind of spatiotemporal dynamics take place in the process where these coexisiting patterns dynamically unstable and develop into spatiotemporal chaos (STC)?

First, the coexisting stable spatial chaos become dynamically unstable and each exhibits different periodic, quasi-periodic and chaotic oscillations. This implies that coexisting orbits are created at a fixed P.

Examples of numerical simulations are shown in Fig. 6.17. Here, the temporal evolution of the local spatial Lyapunov exponent $\bar{\alpha}_L$ averaged over $\Delta/\tau = 5$ is also shown, where

$$\alpha_L(t) \equiv \ln \sum_{k=1}^{N} \delta\phi_k(t)^2/2. \tag{6.40}$$

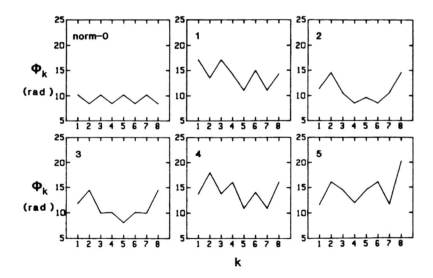

Figure 6.16: Coexisting spatial patterns for $N = 8$. $B = 0.3$, $P = 15$ and $\phi_0 = 0$.

6.2. SPATIOTEMPORAL DYNAMICS

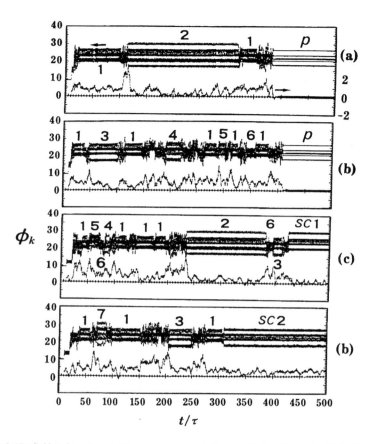

Figure 6.17: Self-induced switching among coexisting spatial patterns in high-P regimes for different initial conditions. p denotes a periodic orbit and sc means *stable* local chaos, from which the system cannot escape. Different attractors among which an itinerant motion takes place are distinguished by numbers. Local spatial lyapunov exponent is also shown (see the right vertical axis).

Here, $\delta\phi_k(t)$ is calculated by the following variational equation

$$\tau\delta\phi_k/dt = -\delta\phi_k + f'_F(\phi_{k-1})\delta\phi_k + f'_B(\phi_{k+1})\delta\phi_{k+1}. \qquad (6.41)$$

Which orbit is searched critically depends on the initial condition. The system searches these orbits wandering chaotically between the ruins of coexisting attractors.

When they are searched, however, the periodic pulsations whose $\bar{\alpha}_L(t) = 0$ servive stably unless the system is perturbed by a strong noise (trigger pulse to a cell, for instance) in the cases of Figs. 6.17(a)–(b). As P

is increased further, the system can hardly find periodic orbits as shown in Fig. 6.17(c), (d) and 'stable' local spatiotemporal chaos (SC) with $\alpha_L > 0$, from which the system cannot escape, is realized. In sufficiently-large-P regions, such ruins of local attractors disappear and the fully developed global STC takes place.

What is the scenario of the self-iduced switching among different spatial patterns. To investigate this problem, let us examine the evolution of on-site field intensities $|E_k|^2 = |E_F^{(k)}|^2 + |E_B^{(k)}|^2$ which provide the on-site nonlinearity giving rise to nonlinear phase rotation. This quantity expresses the phase rotation speed of individual elements, as is seen from Eqs. (6.25)–(6.26). A typical example near the switching point of Fig. 6.17(a) is shown in Fig. 6.18. It is apparent from the figure that electric field of the $k = 4$-th cell, i.e., $|E_4|^2$ decreases at first (see point a in the figure) and this "information" is transferred to adjacent cells successively as shown by arrows

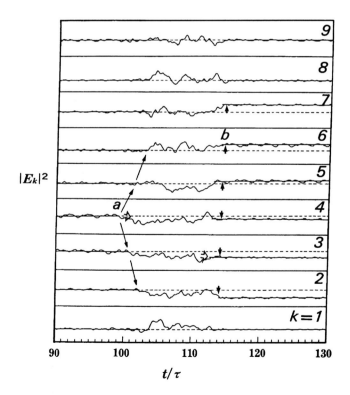

Figure 6.18: Temporal evolutions of field intensities of individual cells near the switching point of Fig. 6.17(a) ($90 < t/\tau < 130$).

due to the mutual coupling between adjacent cells. As a result, all the cells cooperatively excape from the attractor 1. When state values of the *majority* of cells approach the value of a new pattern (i.e., attractor), the system switches to this attractor and self-induced switching to local attractor 2 is established at point b. In any case, particular cell (*intelligent minority*) triggers the switching and other cells follow, leading to a switching to a new pattern. This is the scenario of itinerant behavior in the present system.

6.2.2 Firth model

For a reason of parallel processing, applications of bistable optical elements are likely to involve large numbers of individual elements (pixels) assembled into an array. If one tries to increase the packing density, pixel independence will be destroyed by the cross-talk between neighbors resulting from the mutual couplings. A model of such cross-talk was proposed by Firth and pointed out an interesting connection between chaotic dynamics and pixel independence [14].

A. Basic equation of bistable pixel dynamics

The basic equation models the evolution of a medium parameter ϕ (photocarrier density, atomic inversion etc.) driven by the input optical intensity I_{in}

$$-\ell_D^2 \nabla^2 \phi + \tau \dot{\phi} + \phi = I_{in}(\mathbf{r}, t) f(\phi), \qquad (6.42)$$

where ℓ_D is the transverse diffusion length, τ the excitation decay time. ϕ is generated by I_{in}, and the response function $f(\phi)$ is assumed to be nonlinear in such a way that the constant, uniform response to a plane wave steady input is multivalued, giving optical bistability. The transverse diffusion is the origin of cross-talk in this model.

B. Steady-state analysis

Assume that the input beams are narrow compared to the array spacing and write

$$I_{in}(\mathbf{r}) = \sum_{sites j} P \delta(\mathbf{r} - \mathbf{r_j}). \qquad (6.43)$$

Consider the case where I_{in} excites a linear array: Eq. (6.42) becomes

$$-d^2\phi/dx^2 + \phi = I_{in}(x) f(\phi), \qquad (6.44)$$

in the steady-state. Here, ℓ_D has been scaled out, and I_{in} is periodic in x with period L (times ℓ_D). Equation (6.44) is formally identical to the classical mechanics problem of a nonlinearly driven hyperbolic oscillator. If δ-pumping is assumed as in Eq. (6.43), this becomes a "kicked oscillator" problem, and we can anticipate subharmonic and chaotic solutions, corresponding here to *spatially* unstable spatial patterns. Defining A_n as the value of ϕ on the n-site, the solution is conveniently expressed as a two-dimensional "stroboscopic" area-preserving map:

$$A_{n+1} = 2A_n \cosh L - P\, f(A_n) \sinh L - B_n, \quad B_{n+1} = A_n. \qquad (6.45)$$

This map can be interpreted as a two-term recurrence relation expressing a dependence of the state of the n-th pixel on its nearest neighbors. Figure 6.19 shows iteration of this map for a Lorentzian response function. There are two kinds of fixed points, the middle fixed point surrounded by closed orbits and the hyperbolic fixed point where stable and unstable manifolds intersect. The latter one is very much similar to that discussed in the previous Otsuka-Ikeda bistable chain model. From the linear stability analysis of Eq. (6.42), only hyperbolic fixed points of the map are found to be dynamically stable.

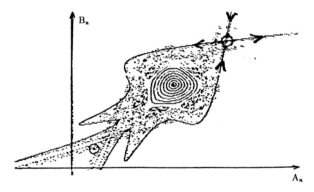

Figure 6.19: Iteration map for a Lorenzitan response function in bistable pixels.

It is clear that most points escape to infinity under the map, which is physically unacceptable. Therefore, one must identify the "bounded set" of points in the (A, B) plane for which A_n remains finite for all n. Firth predicted that the boundary set is a Cantor set, which can be put in 1:1 correspondence with the set of all binary representations of real numbers, for large enough L.

6.3 Cooperative Functions of Coupled Nonlinear Element Systems

In this section, cooperative phenomena in coupled nonlinear element systems presented in the previous sections are described focussing their applicability to signal processing as well as to spatial chaos memory [A], [B], [C].

6.3. COOPERATIVE FUNCTIONS

6.3.1 All-optical signal processing in bistable chain

In unidirectional coupling scheme in the bistable chain system presented in **6.2.1**, dynamically stable spatial structures are restricted to spatially stable periodic solutions, and spatial chaos structures are not realized as stable solutions. However, if the system is operated in a relatively low input intensity regime, it exhibits some interesting cooperative dynamics which can be applied to novel all-optical signal processing, including all kinds of multivibrator operations, flip-flops and a complete set of logic gate operations [C].

A. Domino dynamics in hysteretic regimes

Figure 6.20 shows examples of the input-output characteristics for different cell number N, assuming $B = 0.2$ and $\phi_0 = 0$ in the case of unidirectionally coupled bistable chain. Dashed curves indicate unstable branch. It is clear from this figure that an S-shaped bistable region with a hysteresis appears before the spatial period-doubling takes place.

Assume that $N \gg 1$ and the system is set on the upper state ϕ_u of

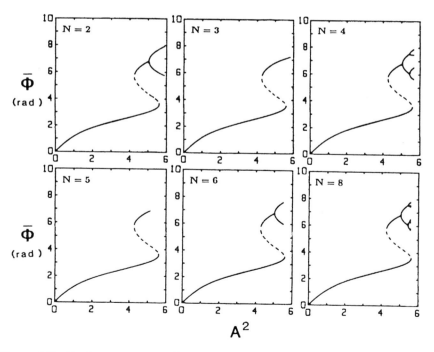

Figure 6.20: Stationary solutions $\bar{\phi}_k$ as a function of input intensity A^2 for uniform excitations in the case of unidirectional interaction. $B = 0.2$ and $\phi_0 = 0$.

S-shaped bistable region and a single cell ($k = 1$) is excited by an optical pulse A_p^2 superimposed on the input bias A^2. Then, ϕ_1 tends to relax to a new "destination"

$$\phi_1^* = f_F^{(1)}(\phi_u) = A_1^2[1 + 2B\eta_F^{(1)}\cos(\phi_u + \phi_0)]$$
$$\equiv A_1^2 g^{(1)}(\phi_u)(A_1^2 = A^2 + A_p^2), \quad (6.46)$$

roughly within a τ period, where $\eta_F^{(1)} \equiv A_F^{(N)}/A_F^{(1)}$. When ϕ_1^* is realized, then the destination toward which the second cell tends to relax is determined by $\phi_2^* = A^2 g^{(2)}(\phi_1^*)$. In this way, ϕ_k of the following cell "falls down" successively in domino fashion toward destinations determined by the mapping rule $\phi_k^* = A^2 g^{(k)}(\phi_{k-1}^*)$. The behavior of destinations of such *domino dynamics* is classified into three cases depending upon how A_p^2 is assigned:
(1) all ϕ_k^* remain in the upper branch (see region A, D in Fig. 6.21(a)),
(2) all ϕ_k^* switch down to the lower branch cooperatively (region B),
(3) all ϕ_k^* go up and down between the upper and lower branches (region C).

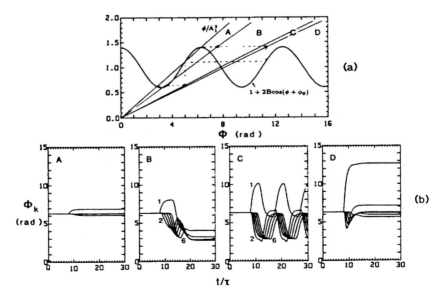

Figure 6.21: Relaxation dynamics when single cell ($k = 1$) is excited. (a) Graphical solution for determining the class of relaxation dynamics, where $g(\phi) \equiv 1 + 2B\cos(\phi + \phi_0)$. (b)Simulated temporal evolution of $\phi_k(t)$ for different regions A, B, C, and D, where the system is initially set on the upper branch of the hysteresis curve. $N = 6$, $B = 0.2$, A^2(bias) = 4.5 and $\phi_0 = 0$. Region A, A_p^2 (trigger pulse intensity to $k = 1$ cell) = 0.5; B, 1.5; C, 3.5; D, 5.5.

6.3. COOPERATIVE FUNCTIONS

Here, $N = 6$, $A^2 = 4.5$, $B = 0.2$ and $\phi_0 = 0$ are assumed. Figure 6.21(b) shows numerical results which confirm three different relaxation dynamics, where a simulation is carried out using Eqs. (6.21) and (6.23) assuming $E_B^{(k)} = 0$ and $\ell/c \ll \tau$. The details of the operation principle of domino dynamics in regions B and C are explained in *Supplement*.

In region C, astable multivibrator operations are possible by applying a pulse to one particular element [C-1]. It is easy to expect that all-optical flip-flops are also possible if the trigger pulse intensity is set in region B or C for on-switchings and is set in region C or D for off-switchings. To achieve the flip-flops, however, there exists a minimum duration time for a trigger pulse. In short, the minimum trigger pulse cutoff timing is determined such that the greater portion of ϕ_k has already crossed over the saddle at the trailing edge of the trigger pulses, as shown in Fig. 6.22. Such a *majority rule* concept is easily understood and is reasonable in terms of domino dynamics. It should be noted that in usual optical bistable device such all-optical flip-flop operations in an S-shaped hysteretic regime are basically impossible since one needs "negative" optical pulses to realize off-switching. In this system, T (Toggle), D (Delay or Data, monostable multivibrator) are also attainable [C-1].

B. Flip-flop and logic-gate operations in pitchfork bifurcation regimes

Above an S-shaped bistable regime in Fig. 6.20, the spatial period-doubling bifurcation takes place. Let us consider the simplest case of $N = 2$. The conceptual model is shown in Fig. 6.23, in which each element is excited by an external laser light and output intensity $|E_m|^2$ ($m = 1, 2$) is extracted from two ports. Figure 6.24(a) shows the stationary solutions as a function of equal laser intensity A^2 to two input ports, assuming $B = 0.5$

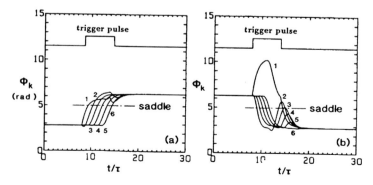

Figure 6.22: All-optical flip-flop operation. A^2(bias) = 4.5, A_p^2 = 3.5, $\Delta t/\tau$(pulsewidth) = 6, $N = 6$, $B = 0.2$, and $\phi_0 = 0$. (a) On switching, (b) Off switching.

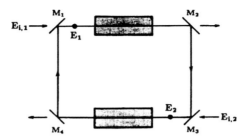

Figure 6.23: Conceptual configuration of a two-element bistable device.

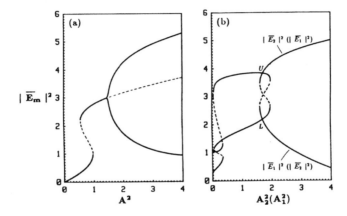

Figure 6.24: Stationary solution versus input intensity.

and $\phi_0 = 4$. Here, unstable regions determined by the linear stability analysis are depicted by dashed curve. It is seen from the figure that the solution above the S-shaped region bifurcates into the asymmetric solution without hysteresis. This is nothing more than a pitchfork bifurcation. Figure 6.24(b) shows output intensity $|E_m|^2$ versus input intensity $A_2^2(A_1^2)$ when $A_1^2(A_2^2)$ is fixed at 1.7, where U and L correspond to the upper and lower state of asymmetric solution at $A^2 = 1.7$ in Fig. 6.24(a).

Let us assume that element I (II) is initially in the lower state L (upper state U). If the trigger optical pulse is applied to element I such that element I makes a transition to the upper state U, element II makes a transition to the lower state L according to the complementally bistable characteristics shown in Fig. 6.24(b). All optical flip-flop operations are realized by utilizing such a switching mechanism. In short, the system is biased in the regime of pitchfork bifurcation and trigger pulses are alter-

6.3. COOPERATIVE FUNCTIONS

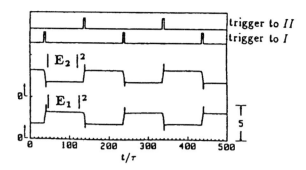

Figure 6.25: All-optical S-R flip-flop operation. The trigger pulse intensity is 1.5.

natively applied to each element. The result of flip-flop operation is shown in Fig. 6.25, assuming $A^2 = 1.7$ (bias), $B = 0.5$ and $\phi_0 = 4$. The intensity and pulse width of the trigger pulses are 1.5 and $\Delta t/\tau = 5$. For element I, the on-switch is realized by the set pulse to I and the off-switch is realized by the reset pulse to II, while element II exhibits complementary behavior. As a result, the set-reset-type flip-flop operation is accomplished on all-optical basis without using "negative" optical pulses similarly to domino dynamics shown in Figs. 6.21 and 6.22.

Next, all-optical logic gates utilizing the complementary output characteristics of Fig. 6.24(b) are shown in Fig. 6.26, where A_1^2 is biased at $A_{1,b}^2 = 1.7$ and input signals $A_{1,s}^2 (\equiv X)$ and $A_{2,s}^2 (\equiv Y)$ are superimposed on the bias input $A_{2,b}^2$.

6.3.2 Spatial chaos memory

If the chaos are realized as dynamically stable spatial structures as described in **6.2**, it can be utilized as spatial chaos memory. Here, we consider the applicability of spatial chaos in the coupled nonlinear element systems to information storage [B].

A. Capacity of spatial chaos memory

As was discussed in the previous section **6.2.1**, spatially unstable chaotic solutions in Otsuka-Ikeda system are dynamically stabilized in the case of bidirectional coupling by the local feedback mechanism. From the viewpoint of memory function, this fact has a very important meaning: The existence of chaotic spatial solutions implies that the number of spatially unstable periodic solutions (with period N) increases exponentially with the number of cells N, i.e., $\exp(hN)$, where h is the topological entropy. In short, the memory capacity of $C = \log h / \log 2$ bit (which is derived

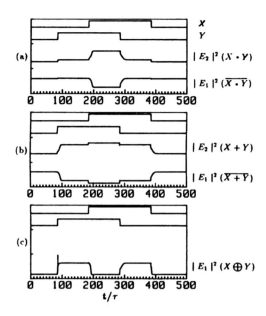

Figure 6.26: Logic gate operations. (a) AND and NAND with $A_{2,b}^2 = 0.4$ (b) OR and NOR with $A_{2,b}^2 = 1.3$ (c) EXCLUSIVE OR with $A_{2,b}^2 = 0$.

from $2^{CN} = \exp(hN)$) is created per unit cell if we assume that all the periodic solutions are dynamically stable. In the present system, however, periodic solutions are not always dynamically stable, as discussed in **6.2.1**. Therefore, $\log h/\log 2$ gives the theoretical upper limit of memory capacity. In Otsuka-Ikeda system, at least all heteroclinic spatial structures which are constructed of arbitrary combinations of α and β structures, are found to be dynamically stable, where α and β structures are formed of five and two cells, respectively. This fact ensures memory capacity of at least 1/7 (bits/cell). That is, the memory capacity of our system C (bits/cell) is evaluated as

$$1/7 \leq C < \log h/\log 2 \quad (bits/cell). \tag{6.47}$$

This value seems to be not so large. However, the important point is that the memory function has been cooperatively created in this system by introducing the bidirectional interaction between cells which do not possess any memory function when the coupling is absent.

B. Assignment to desired spatial chaos patterns

Is it possible to assign the desired spatial patterns to our system? In other words, can spatial chaos memory be addressed easily? If we restrict

6.3. COOPERATIVE FUNCTIONS

the spatial patterns to the structures which are formed of α and β structures, it is quite easy to write any spatial patterns using the following methods.

In one method, first, the trivial period-one cycle structure is realized by setting the system to the low P regime. Then, a weak perturbation which has the same spatial pattern as the desired pattern is applied to the control parameter of each cell and P is increased to the multifurcated region. The linear phase shift $\phi_0^{(k)}$ or the input intensity to each cell $A_F^{(k)2}$ (or $A_B^{(k)2}$) is effective as a control parameter to be modulated. For example, let us consider the case that the linear phase shift is modulated. The seed $\delta\phi_i$ is assumed to be applied to ϕ_0 of each cell, where i is the cell number to be assigned. Then $\delta\phi_i$, which has the α-binary characteristic, is applied to five cells to which an α structure should be assigned. As for β structures, the β-binary characteristic is applied to two cells. Then, P is increased to the multifurcated region. By these processes, the desired structures can be assigned to desired positions. Once the desired patterns are realized, these patterns are frozen even if the perturbation is removed, and they are found to be quite stable against external random noise. Figure 6.27 shows the assigned spatial patterns obtained at different P values.

Let us consider another method. It is shown that period-two cycle

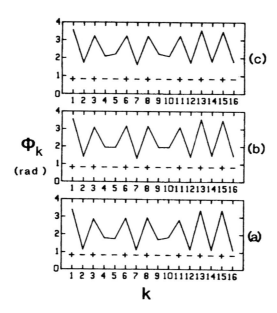

Figure 6.27: Assignment to a desired spatial chaos pattern by modulating the linear phase shift ϕ_0. $N = 16$, $B = 0.3$ and $\phi_0 = 0$. (a) $P = 2.7$, (b) 3.3, (c) 4.0.

structures (for even N) or period-two structures into which a kink is inserted (for odd N) are the most stable "ground-states" in the multifurcated region, i.e., $P > P_c$. It is possible to directly assign the desired spatial patterns consisting of α and β structures. In this method, the "ground state" structures are realized at first by increasing P beyond P_c. Then, a perturbation which has binary characteristics corresponding to the desired spatial patterns is applied to the linear phase shift ($\phi_0^{(k)}$) or to the input intensity ($A_F^{(k)2}$ or $A_B^{(k)2}$) in the form of a *trigger pulse* with a finite pulsewidth. This is very much similar to the assignment to the desired antiphase periodic states by seeding-pulse injection presented in multimode lasers (see **2.2**, **2.3**). This method requires relatively strong perturbations compared to the former method, but, the spatial information input to the trigger pulses can be easily memorized as spatial patterns.

C. Switching between desired patterns: flexibility of spatial chaos memory

Is it possible to switch between spatial chaos patterns by external control? If the answer is yes, we can construct a memory device by summing simple elemental parts. Once the desired spatial patterns are realized, switching to different spatial patterns is easily achieved by applying a perturbation which has the same binary characteristics as the new patterns to the linear phase shift ($\phi_0^{(k)}$) or to the input intensity ($A_F^{(k)2}$, $A_B^{(k)2}$) in the form of trigger pulses as described in the previous subsection *B*. With this perturbation, the old pattern makes a transition to a new one.

The switching process is realized by combining several shifting operations on the α structures along the chain. The elementary process for switching is the shift of a stable kink by two lattice sites as shown in Fig. 6.28(a). In this process, an intermediate structure, i.e., α', appears. This type of kink forms an unstable fixed point, i.e., saddle, of the system and cannot be realized stably [B-2].

Now, let us look at the required switching intensity for shifting operation. The threshold intensity depends upon the duration of the trigger pulse, and has been found to decrease in proportion to $(\Delta t/\tau)^{-1}$ (Δt: pulsewidth). In both cases of the modulation of $A_{F,B}^{(k)2}$ and $\phi_0^{(k)}$, the threshold is as quite small (on the order of 10^{-2}) if the pulsewidth is much longer than the relaxation time of the system τ. Figure 6.28(b) shows the relationship between $\delta\phi_0$ (applied) and the period $\Delta t/\tau$ which is required for the system to cross over the saddle point, where $+\delta\phi_0$ is applied a $k = 10$ cell and $-\delta\phi_0$ is applied to a $k = 11$ cell in Fig. 6.28(a). From Fig. 6.28(b), the barrier height at the saddle point is estimated to be 10^{-3} from $\delta\phi_0$ at $\Delta/\tau \to \infty$ and stable spatial chaos patterns are connected by the extremely flat path through such an extremely low potential barrier [B-2]. This system is distinctive in the sense that the transition between realizable spatial

6.3. COOPERATIVE FUNCTIONS

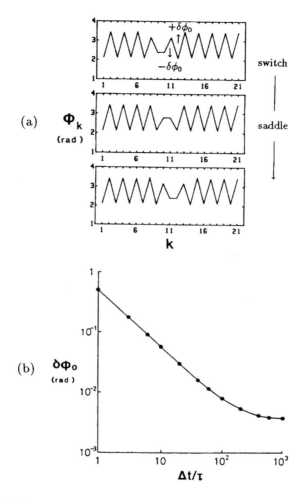

Figure 6.28: (a) Switching among stable "kink" structures via *unstable* kink (e.g., saddle). (b) Relation between the modulation amplitude $\delta\phi_0$ and period $\Delta t/\tau$ which is required for the system to cross over the saddle point in (a). $N = 21$, $P = 5$, $B = 0.3$ and $\phi_0 = 0$.

patterns can be performed quite easily by a slight modulation of the control parameter. In other words, the memory function of our system is rich in *flexibility*.

Figure 6.29 shows an example of the switching process, where perturbations are applied to $A_F^{(k)2}$ in the form of trigger pulses. Re-writable spatial chaos memory is found to be achievable.

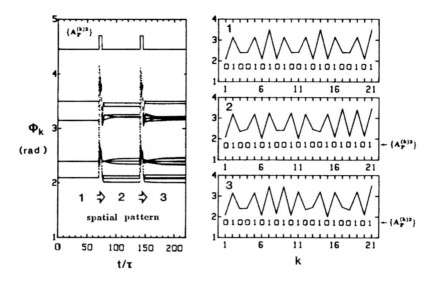

Figure 6.29: Switching between desired patterns indicating a rewritable spatial chaos memory for $P = 5$. Input signals $A_p^{(k)2}(k = 1, 2, \ldots, N)$ having binary characteristics is superimposed on the bias input $A_{F,b}^2 = A_{B,b}^2 = 2.5$. In the binary characteristics shown in (b), "1" means $A_p^{(k)2} = 0.5$ and "0" means $A_p^{(k)2} = 0$. $N = 21$, $B = 0.3$, and $\phi_0 = 0$.

The important point of spatial chaos memory described in this section is that it satisfies the minimum demand for information storage functions, although the memory capacity seems to be not so large if the discussion is restricted to the heteroclinic chaos. In the spatial chaos memory, the information storage is accomplished by spatial chaos patterns which are formed by the cooperative interaction between individual elements rather than by individual elements themselves. The memory function acquired by such a mechanism is expected to possess novel properties (e.g. flexibility). The stabilization of spatial chaos by bidirectional interaction, i.e., local feedback mechanism, may provide a support with the stable existence of spatial polarization chaos in four-wave mixing scheme descussed in **6.1.2**. If this is the case, nonlinear polarization dynamics will find an application to information storage. To realize complex functions by applying the distinctive characteristics of chaos, the dynamic aspect of spatial chaos of more realistic optical systems should be investigated extensively.

D. *Spatial chaos memory operation in Firth model*

Homoclinc orbits in Firth model (**3.2.2**) are closely associated with spatial chaos as is discussed in Otsuka-Ikeda model (**3.2.1**), in which symbolic terms means that the dynamics possesses orbits describable by arbitrary

6.3. COOPERATIVE FUNCTIONS

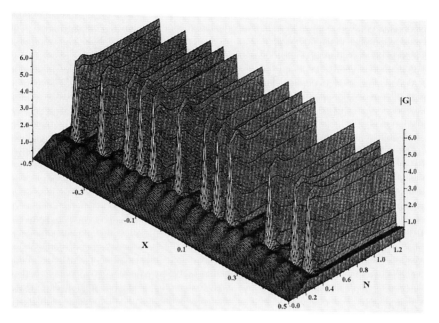

Figure 6.30: Spatial chaos memory in coupled bistable pixels. A small modulation on the backgrounding Gaussian beam defines pixel sites in the spatial dimension (X). Output field, $|G|$, is shown during initialization of the hold beam and the rapid encoding of arbitrary bit sequence. The entire simulation is over 130 cavity transits.

strings of binary digits. If these chaotic spatial patterns are dynamically stable, the system acts as a memory device. Indeed, Firth demonstrated the assignment to desired chaotic patterns by slightly modulating the phase of background Gaussian beam spatially. Figure 6.30 is an example of results of numerical simulation [16].

Supplement: Principle of Domino Dynamics

(1) Cooperative switching

As mentioned in **6.3.1 A**, the input field to a specific cell should be slightly changed from the input field to other cells in order to realize the domino dynamics. This introduces the following cell dependence of coupling function:

$$f_F^{(k)}(\phi) = A_F^{(k)2}[1 + 2B\eta_F^{(k)} \cos(\phi + \phi_0)] \tag{S.1}$$

via the two terms $A_F^{(k)2}$ and $\eta_F^{(k)} \equiv A_F^{(k-1)}/A_F^{(k)}$. The modulation due to the former term is essential for the operation of domino dynamics. To avoid nonessential complexity, we approximate $\eta_F^{(k)}$ by 1 even if $A_F^{(k)}$ is modulated. Then, the coupling function is expressed as $f_F^{(k)} \equiv A_k^2 g(\phi)$, where $g(\phi) = 1 + 2B\cos(\phi + \phi_0)$.

Assume that $N \gg 1$ and that all the cell are set on the upper branch of period-1 cycle structure in a hysteretic regime, which is determined by $\phi_u = A^2 g(\phi_u)$. When the input light intensity of the first cell ($k = 1$) is increased such that $A^2 \to A_1^2$ (A_1^2 is within B regime of Fig. 6.21 (a)), the destination of relaxation of ϕ_1 is changed from ϕ_u to $\phi_1^* = A_1^2 g(\phi_u)$. [See Fig. S(a).] As a result, ϕ_1 relaxes to ϕ_1^* and increases once as seen in Fig. 21(b).

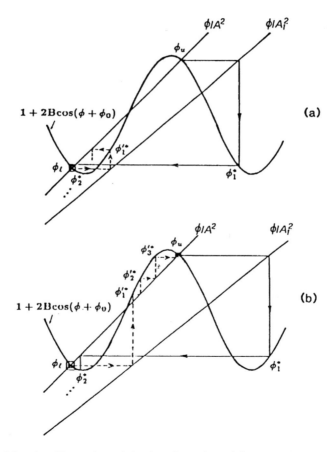

Figure S: Mapping illustration of domino dynamics. (a) cooperative switching, (b) astable multivibrator operations.

However, since ϕ_1^* belongs to the basin of attraction of the lower branch, the destination toward which the following cells tend to relax converges to the lower branch solution ϕ_ℓ such that $\phi_2^* = A^2 g(\phi_1^*), \phi_3^* = A^2 g(\phi_2^*), \ldots$, as shown in Fig. S(b). Therefore, ϕ_N tends to relax ϕ_N^*, which is very close to ϕ_ℓ, and the destination of relaxation of ϕ_1 is then changed from ϕ_1^* to $\phi_1^{'*} = A_1^2 g(\phi_\ell)$. Since $\phi_1^{'*}$ belongs to the basin of attraction of the lower branch [Fig. S(a)] by the definition of region B, all ϕ's are frozen to the stable spatial pattern $\bar{\phi}_1 = A_1^2 g(\bar{\phi}_N), \bar{\phi}_2 = A^2 g(\bar{\phi}_1), \ldots, \bar{\phi}_N = A^2 g(\bar{\phi}_{N-1})$ and cooperative switching is realized.

(2) Multivibrator operation

If we choose A_1^2 within the C regime of Fig. 6.21(a), all behavior in the first half is the same as that of (1), and the destinations of relaxation are changed to $\phi_1^*, \phi_2^*, \ldots$. As a result, ϕ_1 increases once as seen Fig. 6.21(b) and $\phi_2, \phi_3, \ldots, \phi_N$ relax to ϕ_ℓ. The different dynamics arise in the latter half. Due to the definition of region C, the new destination of relaxation $\phi_1^{'*} = A_1^2 g(\phi_N) \simeq A_1^2 g(\phi_\ell)$ belongs to the basin of attraction of the upper branch [Fig. S(b)]. Therefore, new destinations of $\phi_2, \phi_3, \ldots, \phi_N$ are switched to the upper branch such that $\phi_2^{'*} = A^2 g(\phi_1^{'*}), \phi_3^{'*} = A^2 g(\phi_2^{'*}), \ldots$, and relax susccessively toward ϕ_u as shown in Fig. S(b). The same process occurs repeatedly hereafter and astable multivibrator operation takes place.

As emphasized in the beginning of this *supplement*, we used the approximation $\eta_F^{(k)} \simeq 1$ throughout the above considerations. The effect of deviation of $\eta_F^{(k)}$ from 1 is by no means negligible, especially when B is very small (say, $B < 0.1$). In this limit, the modulation effect reduces region B, thereby widening region C. However, the modulation effect is not significant if B is not very small and the above mentioned considerations are substantially correct.

Keynote Papers

This Chaper is written on the basis of the following papers:
[A]: Section 6.1
1. J. Yumoto and K. Otsuka, "Frustrated optical instabilities: self-induced periodic and chaotic spatial distribution of polarization in nonlinear optical media", *Phys. Rev. Lett.*, Vol. 54 (1985) 1806–1809.
2. K. Otsuka anf J. Yumoto, "Polarization bistability and instability in degenerate four-wave mixing", *SPIE Proc. on Optical Chaos*, Vol. 667 (1986) 167–174.
[B]: Section 6.2
1. K. Otsuka and K. Ikeda, "Self-induced spatial disorder in a nonlinear optical system", *Phys. Rev. Lett.*, Vol. 59 (1987) 194–197.
2. K. Otsuka and K. Ikeda, "Cooperative dynamics and functions in a collective nonlinear optical element system", *Phys. Rev. A*, Vol. 39 (1989) 5209–5228.
3. K. Otsuka, "Chaotic itinerancy in a coupled-element multistable optical chain", *Phys. Rev. A*, Vol. 43 (1990) 618–621.
[C]: Section 6.3

1. K. Otsuka and K. Ikeda, "Hierarchical multistability and cooperative flip-flop operation in a bistable optical system with distributed nonlinear elements", *Opt. Lett.*, Vol. 12 (1987) 599–601.
2. K. Otsuka, "All-optical flip-flop operation in a coupled element bistable device", *Electron. Lett.*, Vol. 24 (1988) 800–801.
3. K. Otsuka, "Pitchfork bifurcation and all-optical digital signal processing with a coupled-element bistable system", *Opt. Lett.*, Vol. 14 (1989) 72–74.

Other nonlinear functions and dynamic devices are proposed in the following papers:
a. K. Otsuka, "Nonlinear Antiresonant Ring Interferometer", Opt. Lett., Vol. 8 (1983) 471–473. [This is the first key proposal which gave a birth of *nonlinear loop mirror* being widely used for all-optical high-speed signal processing.]
b. K. Otsuka, J. Yumoto, and J. J. Song, "Optical bistability based on self-induced polarization-state change in anisotropic Kerr-like media", *Opt. Lett.*, Vol. 10 (1985) pp. 508–510.
c. K. Otsuka, "Pitchfork bifurcation and all-optical flip-flop operation with a coupled traveling-wave amplifier system", *J. Opt. Soc. Am.*, Vol. B8 (1991) pp. 1304–1306.
d. K. Otsuka, J.-L. Chern, and J. McIver, "High speed self-pulsation in a two-element optical bistable device", *Opt. Commun.*, Vol. 76 (1990) pp. 245–249.

References

1) Hopfield, Proc. Natl. Acad. Sci. U.S.A. **79** (1982) 2554.

2) Y. Owechko, E. Marcom, B. H. Soffer, and G. Dunning, SPIE 700-1986 International Optical Computing Conference, (1986) 296.

3) P. D. Maker, R. W. Terhune, and C. M. Savage, Phys. Rev. Lett. **12** (1964) 507.

4) C. C. Shang and H. Hsu, IEEE J. Quantum Electron. **QE-23** (1987) 177.

5) G. Gregori and S. Wabnitz, Phys. Rev. Lett. **56** (1986) 600.

6) S. Wabnitz and G. Gregori, Optics Commun. **59** (1986) 72.

7) S. Wabnitz, Phys. Rev. Lett. **58** (1987) 1415.

8) M. V. Tratnik and J. E. Sipe, Phys. Rev. A **35** (1987) 2976.

9) L. W. Pars, *Analytical Dynamics* (Heinemann, London, 1965).

10) M. V. Tratnik and J. E. Sipe, Phys. Rev. A **36**, 4817 (1987).

11) A. E. Kaplan and C. T. Law, IEEE J. Quantum Electron. **QE-21** (1985) 1529; J. Yumoto and K. Otsuka, Phys. Rev. A **34** (1986) 4445; M. I. Dykman, Sov. Phys. JETP **64** (1986) 927.

12) D. J. Gauthier, M. S. Malcuit, A. L. Gaeta, and R. W. Boyd, Phys. Rev. Lett. **64** (1990) 1721.

13) Aubry, in *Solitons and Condensed Matter Physics*, edited by A. R. Bishop and T. Schneider (Springer-Verlag, Berlin, 1979), p. 264.

14) W. J. Firth, Phys. Lett. A **125** (1987) 375.

15) J. Guckenheimer and P. Holmes, *Nonlinear Oscillations, Dynamical Systems, and Bifurcations of Vector Fields* (Springer-Verlag, Heiderberg, 1983).

16) G. S. McDonald and W. J. Firth, *OSA Topical Meeting on Nonlinear Dynamics in Optical Systems* Afton, OK (1990).

INDEX

α parameter, 22, 23
$\chi^{(3)}$ nonlinearity, 3, 23, 251
Λ-scheme, 195, 197
π out-of-phase mode, 224

absorption length, 186, 191
adiabatic elimination, 66, 68
Allan variance, 158, 161, 165
AM-locking, 229
amplified spontaneous emission, 33
amplitude-phase coupling, 225
anomalous dispersion effect, 22
antimodes, 37
antiphase dynamics, 119, 127, 132, 170, 188, 189
antiphase intracavity second-harmonic generation, 88, 99
antiphase periodic state, 67, 68, 89, 91,
antiphase selfpulsation, 195, 197
area-preserving map, 276
associative memory, 249
attractor crowding, 67
attractor ruins, 36, 88, 226
Auger parameter, 192
Auger recombination process, 157, 185, 191

bad cavity condition, 6, 18
band structure, 23
basin of attraction, 229
Belousov-Zhabotinskii reactions, 208
Bernoulli map, 165
bifurcation diagram, 69, 72, 73, 90, 172, 237, 264
bifurcation tree, 45
birefringent optical fiber, 255

bistability, 42, 174, 233
bistable chain, 277
bistable pixel, 261, 275
Bloch equation, 14
breathing motions, 135, 139

Cantor set, 276
Casperson instability, 13
central limit theorem, 162
chaotic burst, 175, 181
chaotic itinerancy, 36, 63, 64, 88, 93, 116, 218, 223, 226, 228, 229
chaotic relaxation oscillation, 69, 130, 175, 179
chaotic search, 46
chaotic switching, 102
characteristic equation, 234, 240, 264, 267
circulation analysis, 91, 101, 133
circulation, 72, 91, 101, 133
class-A laser, 24, 201
class-B laser, 22, 66, 119, 158, 196, 219
class-C laser, 22
clustered state, 69, 84, 102, 128, 132, 139, 241
coherence collapse, 36
collective behavior, 249
combination-tone polarizations, 26
complex Ginzburg-Landau attractor, 203
complex Ginzburg-Landau chaos, 208
compound-cavity eigenmode, 29, 175
controlling chaos, 169
cooperative intensity transfer, 95
cooperative synchronization, 78
core mode, 63
core region, 61
coupled chemical oscillators, 67

coupled generalized Van der Pol laser equations, 35, 47
coupled Josephson junctions, 67
coupled-waveguide lasers, 219
creative minority, 218
crises, 205, 237
cross saturation, 69, 89, 146
cross-talk, 275
crystal optics, 250

damping rate of the optical cavity, 2
Debye relaxation equation, 251
defect, 107, 114, 266
degenerate four-wave mixing, 257, 261
delay-differential equation, 37
density matrix elements, 2
devil's staircase, 27
diffusion distance, 191
dispersion, 3
dispersive limit, 39
distributed feedback LD, 245
distributed nonlinear passive optical elements, 249
domino dynamics, 277, 278, 287
double combs, 33
Duffing oscillator, 253, 256, 258
dynamic stability, 265, 267
dynamical equipartition, 72, 82
dynamical memory, 42

easy switching paths, 86, 99
eigenpolarizations, 254, 255
electric dipole moment, 2
electromagnetic field, 2
ellipse rotation, 250, 254
emitter partition noise, 219
energy conservation condition, 40

factorial dynamic memory, 67, 84, 116, 231
factorial dynamic patterns, 113
far-field pattern, 223, 225
far-infrared lasers, 11
Faraday rotator, 245
Farely fractions, 27
field continuum condition, 33
Firth model, 275, 286
flip-flop, 277, 279

FM-laser operation, 24
focus, 203
four-wave mixing, 23, 39
fractional locking, 35
free-carrier plasma effect, 22
Frenkel-Kontrova system, 261
frustration, 157, 175, 176, 178

gain circulations, 72
Gaussian Markov property, 162, 167
generalized bistability, 174, 233
Gestalt, 249
gigabit picosecond pulse generation, 116, 245
global chaos, 59, 93
global intensity circulation, 101
global Lyapunov exponent, 160
globally-coupled laser arrays, 54, 230
globally-coupled multimode laser, 54, 66
green problem, 187
grouping chaos, 67, 78
grouping state, 76, 102, 103

hard mode, 236, 237
Henon map, 161
Hermite-Gaussian oscillation modes, 145
heteroclinic chaos, 205, 269
heteroclinic connections, 115
heteroclinic crossing, 270
heteroclinic orbit, 269
high-density pumping, 185, 190
hole solution, 215
homoclinic attractors, 216
homoclinic chaos, 205, 208
homogeneous linewidth, 54
homogeneously-broadened single-mode laser, 3
homogeneously-broadened multimode laser, 54
Hopf bifurcation, 5, 14, 88, 91, 196, 209
Hopfield model, 201
hybrid bistable device, 49
hybrid optical system, 46
hyperbolic fixed point, 8, 270
hysteresis, 266

Ikeda instability, 39
Ikeda map, 42

Index 293

in-phase mode, 224
incoherent delayed feedback, 158, 233
incoherent feedback, 230
inhomogeneous broadening, 14
inhomogeneous linewidth, 54
inhomogeneously broadened single-mode laser, 13
injection locking, 34
injection seeding, 68, 71, 76, 82, 116
intensity circulation, 82, 91
intermittency, 22
intermode statistical non-independence, 130
inverse power-law relation, 73, 86
inverse type-II intermittency, 36
isomers, 42, 45

Jacobian elliptic functions, 253
Jacobian matrix, 4, 56

Kerr medium, 251
kicked oscillator, 275
kink, 32, 270
Kolomogorov-Arnol'd-Moser (KAM) tori, 257
Kramers-Kronig relationship, 3, 23

large-capacity optical transmissions, 116
laser array, 201
laser complex systems, 53
laser Lorenz chaos, 11
laser-diode arrays, 224
laser-diode-pumped microchip laser, 154, 185
$LiNdP_4O_{12}$ (LNP) laser, 124, 128, 135, 145, 152, 158, 169, 175, 186
linear stability analysis, 3, 122, 234, 240, 264
local chaos, 59, 63, 93
local Lyapunov exponent, 158, 160, 164, 166
local spatial Lyapunov exponent, 272
logic gate, 277, 279
longitudinal mode spacing frequency, 1
longitudinal relaxation rate, 1
Lorenz chaos, 208
Lorenz Equation, 7
Lorenz-Haken instability, 6

Lyapunov exponent, 77, 80, 111, 158, 164
Lyapunov functional, 265
Lyapunov spectrum, 77, 79, 94, 106

majority rule, 212, 216, 229, 279
material polarization, 2
Maxwell-Bloch equations, 4, 54
Maxwell-Bloch turbulence, 54
McCumber frequency, 123, 189
memory capacity, 45, 116, 282
mode hopping, 63
mode partition noise, 62, 175, 181, 182
mode-locking, 24, 27
mode-splitting phenomenon, 15
modulated multimode lasers, 68, 127
multichannel self-mixing laser-Doppler-velocimetry, 137, 169
multifurcation, 265
multiple-parametric resonances, 138
multiple-transition oscillations, 188
multivibrator, 277, 289

Navier-Stokes equation, 7
near-subharmonic resonance, 174
nearly degenerate four-wave mixing, 24
nearly resonant perturbation, 157
neural networks, 249
node, 203
nonequilibrium systems, 250
nonlinear eigenpolarizations, 259
nonlinear loop mirror, 290
nonlinear passive resonator, 37
nonlinear polarization dynamics, 250
nonreciprocal independence, 82
nonstationary chaos, 162, 167
nonstationary-to-stationary transition of chaos, 157
norm, 228
number-theoretic method, 175

OGY method, 169
olfactory sensing systems, 67
optical bistability, 275
orthogonal polarization, 89
Otsuka-Ikeda model, 262

pair annihilation, 191
parametric four-wave mixing, 33

parametric gain, 32
parametric resonance, 135, 139
parity-invariant crystal, 252
period-doubling, 13, 20, 22, 37, 39, 130, 169, 203, 209, 238,
phase hopping, 219, 227, 228
phase inflection, 62
phase locking, 24, 222, 229
phase rotor, 209, 213, 216, 229
photon lifetime, 68
pitchfork bifurcation, 279
Poincaré map, 11, 257
Poincaré plot, 36, 261
Poincaré section, 240
Poincaré sphere, 253
pony on the merry-go-round, 185
population grating, 196
population inversion, 68, 119
population lifetime, 68
population pulsation, 14, 17, 24
potential barrier, 284
power spectral density, 165
pump induced coherence, 12

Q-switching antiphase periodic states, 99
quadratic-to-quartic transition, 157, 192, 193
quantum mechanical harmonic oscillator, 229
quasiperiodicity, 22, 27, 31, 93, 259

Rabi force, 61, 66, 223
Rabi frequency, 55
Rabi nutation, 10
Rabi precession frequency, 1
Rabi precession, 5, 7, 11, 16, 53, 57
Rabi unstable bands, 61, 63
reabsorption process, 191
relaxation oscillation frequency, 1, 122, 135, 145, 166, 169
resonance state, 259
resonant Rabi instability, 54, 57
resonant-type bistable laser-diode amplifier, 42
return map, 11, 78, 270

saddle, 203, 254, 271, 284
Schrödinger equation, 2

second threshold, 5, 17, 57
second-order nonlinear process, 88
seeded bifurcation switch, 46
self-induced phase turbulence, 218, 221, 223
self-mixing laser-Doppler velocimetry, 128, 148, 169
self-mode-locking, 58
self-organized criticality, 158
self-organized relaxation oscillations, 120
self-sustained relaxation oscillation, 28, 29
semiclassical equations, 2
semiconductor laser, 22, 63, 116
separatrix orbit, 254, 255
Shannon's mutual information, 66
Shil'nikov-type orbit, 208
shot noise, 182
single mode stationary solution, 55
slow manifold, 61
soft mode, 235, 240
spatial bifurcation phenomena, 255
spatial chaos memory, 281, 286
spatial chaos, 250, 255, 258, 261, 264, 265
spatial disorder, 250, 265
spatial harmonic oscillator, 250
spatial heterostructure, 269
spatial hole-burning, 66, 68, 119, 185, 191
spatial instability, 267
spatial Lyapunov exponent, 265
spatial period-doubling, 264, 277
spatiotemporal chaos, 208, 226, 265, 271, 274
spectral hole-burning, 15
spiking laser oscillations, 157
spiking-mode oscillation, 68, 116, 236
spontaneous emission coefficient, 68
spot dancing, 221, 223, 225
stability exponents, 8, 270
stable manifold, 9, 269
stagnant motion, 84
stagnation phenomena, 228
standard deviation, 161, 164
stationary chaos, 161
Stokes parameters, 252
Stokes vector, 252
strange attractor, 157, 158
subharmonic resonance, 24, 32, 47
sum rule, 111

super-slow relaxation, 219, 226
supermodes, 226, 227, 228
suppression of chaos, 171
surface-emitting LD, 245
sustained relaxation oscillation, 30, 69, 103, 180
switching-path formation, 86, 112, 216
symmetry-breaking, 203, 206, 209
synchronized pulsation, 241
synchronized relaxation oscillations, 130

Tang-Statz-deMars (TSD) equations, 120, 142, 173
the least common frequency, 243
thermal convection equation, 7
third-order nonlinearity, 28
third-order susceptibility tensor, 252
three-dimensional self-organization, 144, 150
3-point mutual information, 66
time-dependent complex Ginzburg-Landau equations, 202

Toda oscillator, 234
Toda potential, 234
topological entropy, 281
torus, 60, 61, 239
transient chaotic itinerancy, 84
transverse clustering, 149
transverse effects, 144
transverse relaxation rate, 1
transverse synchronization, 148
tree-structure, 42

undulation curve, 30
uniaxial crystal, 252
unstable manifold, 8, 9, 269
unstable Rabi bands, 57

vanishing gain circulation rule, 72, 77, 82, 134
variational equation, 269, 273

winding number, 27, 63, 240
winner-takes-all dynamics, 185